FORESTRY COMMISSION
FIELD BOOK 8

The Use of Herbicides in the Forest – 1989

by **D. R. Williamson**
Silviculturist, Forest Research Station, Alice Holt Lodge, Wrecclesham, Farnham, Surrey GU10 4LH

and **P. B. Lane**
Forestry Commission Work Study Team, Eastern Region, Santon Downham, Brandon, Suffolk IP27 0TJ

London: Her Majesty's Stationery Office

© *Crown copyright 1989*
First published 1983
Third edition 1989

ISBN 0 11 710270 9
ODC 414:441:236.1:307

Keywords: *Herbicides, Forestry*

Enquiries relating to this publication
should be addressed to:
the Technical Publications Officer,
Forestry Commission, Forest Research Station,
Alice Holt Lodge, Wrecclesham,
Farnham, Surrey GU10 4LH

Further copies of the wallchart
Guide to the use of herbicides in the forest
are obtainable free of charge from the above address.

Contents

Preface

This Field Book has been prepared as a successor to
Forestry Commission Booklet 51 (1986 edition) *The use of
herbicides in the forest*. Since that Booklet was published,
two sets of statutory regulations have come into force
which materially alter the framework in which any pesticide
(i.e. herbicide, fungicide, insecticide, etc.) is used. These are
The Control of Pesticides Regulations 1986 and The
Control of Substances Hazardous to Health Regulations
1988.

The main difference between the content of this Field
Book and its Booklet predecessor is that present recommendations are linked to proprietary products, approved for
the uses specified on the product label, whereas previously,
recommendations were in terms of the active ingredient.
Formerly, the potential user had then to determine which
proprietary product was available on the market.

All recommendations made in this Field Book are
believed to be supported by valid 'on-label' or 'off-label'
approvals under current regulations.

Linking recommendations to products carries with it the
risk that manufacturers may, for commercial reasons, with-
draw a product from the market at short notice. While in
principle users can apply for 'off-label' approval to enable
them to use an alternative, in practice the delays currently
being experienced in examination of all applications for
approvals, whether 'off-label' or 'on-label', extend to well
over 12 months. Consequently, if weeds or other pests need
urgent treatment, a prescription for treatment has to be
sought among the products which have an existing
approval.

Weedkillers, responsibly used, have been of great
assistance in containing the costs of establishing or re-
establishing forest crops. This Field Book should provide
the basis for a successful continuance of this practice.

1 Introduction

1.1 *General information*

Changes in the legislation surrounding the use of pesticides
have led to the preparation of this Field Book which
replaces Forestry Commission Booklet 51.

All recommendations within this Field Book conform to
present legislation under the Control of Pesticides
Regulations 1986 and the Control of Substances Hazardous
to Health Regulations 1988 and to the *Code of practice for
the use of pesticides in forestry* (pesticides include
herbicides, insecticides, fungicides, etc.)

This Field Book contains recommendations for the use of
herbicides in the forest. In this context, forestry includes
the establishment of trees in new plantations, restocking of
forests and woodlands, shelterbelts and ride side edges.
Control of weeds around trees in parks and gardens is not
covered by these recommendations.

The products recommended in this Field Book are those
of which the Forestry Commission Research Division has
experience. These should prove adequate in most situations,
but not all herbicides with label approval for use in forestry
or amenity tree situations are listed. The omission of particular products does not necessarily mean that they will be
less effective: for example, some products have been
omitted because there is a safer alternative available.

The application equipment described in this Field Book
is a representative sample of the various applicators which
the Forestry Commission's Work Study Branch have found
to be effective.

Since the introduction of the Control of Pesticides
Regulations 1986, legal approval for the use is given only
to specific manufacturers' products.

A major requirement of these new regulations is that the
users must comply with the main conditions of approval as
stated on the label or in an 'off-label approval' (see Section
2). The Control of Substances Hazardous to Health
Regulations 1988 reinforce provisions to ensure the operators and others involved in any pesticide use are exposed

1

as little as is practicable.

All the recommendations given in this Field Book are covered by either:

a. product label approval;
b. extensions of label approval allowable under the *Code of practice for the use of pesticides in forestry* (see Section 2);
c. an application for 'off-label approval' (see Section 2), which is expected to be granted before the end of 1989.

Any pesticide used under an 'off-label approval' is at user's risk.

This Field Book will be revised as frequently as is necessary to keep it up to date and useful as a working manual. Between editions updating information will be made available through Research Information Notes (RINs), published by the Forestry Commission, Forest Research Station, Alice Holt Lodge, Wrecclesham, Farnham, Surrey GU10 4LH.

Recommendations for weed control in forest nursery seedbeds and transplant lines will be included in a publication (in preparation) that will replace Forestry Commission Booklet 52, now out-of-print.

1.2 *Content and layout*

The main sections of this Field Book (Sections 4–9) are laid out by reference to major weed types: grass/herbaceous broadleaved, bracken, heather, woody weeds, gorse and broom, and rhododendron. The herbicides appropriate for use against this range of forest weed species are set out in the accompanying wallchart. Once candidate herbicides have been selected, detailed information on particular products can be obtained from the relevant section of the Field Book.

Before making a final selection of any herbicide, managers must consider whether there are other more appropriate means of achieving their objectives than by use of a herbicide.

Herbicide entries are set out in a standard format:
  description and properties of approved products,
  crop tolerance,

1

recommended application rates,
methods and timing of application,
additional information on weed control,
protective clothing and special precautions.

Sections 2 and 3 on legislation and safe working practices and Sections 10 and 11 on protective clothing and
application equipment contain very important information
upon which the success of any herbicide treatment depends.
PLEASE READ THEM CAREFULLY AND FOLLOW
THEIR GUIDANCE whenever you undertake a
programme of weed control with herbicides.

1.3 Nomenclature of herbicides

Herbicides are from time to time referred to by one of the
following different types of name: e.g. for glyphosate,

Product name: Roundup—the name registered by the
                  manufacture for a product containing a
                  specific formulation of an active ingredient
                  (a.i.). Approval is sought and given for the
                  product by name.

Common name: Glyphosate—the accepted short name for
                  an active ingredient.

Chemical name: N-phosphonomethyl glycine; the full
                  scientific name for the active ingredient.

These terms could be combined as follows: Roundup is a
liquid formulation of glyphosate, containing 360 grams per
litre (g/l) of an active ingredient (N-phosphonomethyl
glycine) as the mono isopropylamine salt.

In this Field Book only the product name or common
name is used.

1.4 Assumptions and conventions used in this Field Book

(see also Section 13 for glossary and list of abbreviations)

1.4.1 Area

Throughout this Field Book, unless the context clearly indi-
cates otherwise, all references to areas (usually hectares)
refer to TREATED AREA, that is the area of ground or

plantation that is actually covered with herbicide, i.e. the total area of spots or bands treated within a plantation.

1.4.2 Crop tolerance

The descriptions of crop tolerance assume average site conditions and healthy crop trees (prior to treatment). Not only is the crop at risk if an overdose is applied but under the Control of Pesticides Regulations, it is illegal to apply a rate of pesticide greater than that stated on the label for the use of that pesticide.

For each herbicide entry, where appropriate, there is a note of any necessary waiting period after treatment with the herbicide and before it is safe to plant on the site.

The manager should bear in mind that crop tolerance and weed susceptibility can be affected by site, crop condition, provenance, weather, season, etc. He should proceed with caution in choosing herbicides and rates until he is confident of the efficacy of treatment in the local conditions within which he is working.

1.5 Application patterns

A herbicide can be applied in three ways:

overall—the herbicide is applied over the whole weeding site;

band —the herbicide is applied in a band over or between crop trees;

spot —the herbicide is applied as individual spots around each tree.

A directed application is a spray which is directed to hit a target pest or weed and to avoid crop trees.

It is usually easier and quicker, where terrain and crop allow, to apply herbicide as an overall treatment, but this economy of effort imposes a greater risk to the local woodland habitat and uses more herbicide than is strictly required for effective weed control.

When planning pre-planting weed control it is important to consider whether planters will be able to identify treated areas if herbicides are applied in spots or bands. These two techniques may be inappropriate if rhizomatous or stoloniferous weeds such as couch grass are present.

In a post-planting situation where overall application is undesirable spot or band treatment may be appropriate because usually young trees can be clearly seen.

Directed spot application patterns are more demanding of time and skill but minimise the amount of herbicide used, the risk to which the crop is exposed and the impact on the environment. It is easily carried out where trees are enclosed in treeshelters.

1.6 Application methods

The equipment used to apply a herbicide will depend upon the nature of the herbicide (granular or liquid) and on the application pattern required.

Factors to consider when contemplating:

Granular application
a. no water is required
b. no mixing required
c. low cost applicators
d. no problems associated with the disposal of unused diluted pesticide
e. granular applicators are difficult to calibrate accurately and it is therefore more difficult to achieve the correct application rate per spot
f. granular products are often more expensive
g. granular products are bulkier to store and transport
h. limited range of products.

Liquid application
a. water is required except with a few ULV (ultra low volume) products
b. mixing is required except with a few ULV products
c. applicators are more expensive than the simple granular applicators
d. applicator failure or climatic change can lead to the need to dispose of unused diluted pesticide
e. liquid applicators can be calibrated accurately (calibration of the Forestry Spot Gun is particularly simple)
f. there is a greater choice of liquid products
g. with many products the volume rate of application can be altered to suit various types of applicator

1

h. liquid or wettable powder formulations are less bulky to store and transport.

Section 11 gives detailed guidance on the various types of applicator available.

Aerial application of herbicides is covered by the Control of Pesticides Regulations 1986. Managers wishing to consider this technique must first study the legal obligations under these regulations and should then contact the manufacturer of the product if they wish to consider aerial application further. Only one herbicide (Asulox) is currently fully approved for aerial application; even so, users must comply with the regulations.

1.7 *The decision chain—a summary*

The following sequence briefly describes the assessments and decisions involved in achieving correct application of a liquid herbicide. The sequence for a granular herbicide is similar but with rather fewer variables to reconcile.

Using Sections 4 to 9:

a. From crop and weed characteristics, determine the choice of suitable herbicides, dose rates and application patterns.

b. Make an assessment of the effects of the proposals on operators and the environment ensuring that all necessary equipment is available, safeguards known and competent and certificated operators available.

c. Consider specifically whether any factors of the locality, operation, etc., add to the risk normally associated with the proposed use.

d. Consider any other factors limiting the choice of herbicide, applicator, droplet size, dilution rate or diluent.

e. Select a suitable:

herbicide
dose rate
applicator
droplet size (if critical)
application pattern and method
application rate.

1

f. **Using Section 10:**
Having decided on the herbicide and method of application, ensure suitable protective equipment is available to operators.

g. **Using Section 11:**
i. Calculate the likely equipment requirements and settings, choosing values for the relevant variables, e.g. for the knapsack sprayer:
walking speed
nozzle size
swathe width
application rate
dilution.
ii. Calibrate the equipment to achieve the correct application rate.

1.8 *References for further reading*

Several references to other published books and leaflets are to be found in the appropriate sections of this Field Book. The following titles will also be useful for background reading and to provide further details of some aspects of herbicide practice:

Forestry Commission publications

Bulletin 73: *Rhododendron ponticum as a forest weed,* 1987.
Handbook 2: *Trees and weeds—weed control for successful tree establishment,* 1987.

Occasional Paper 21: *Provisional code of practice for the use of pesticides in forestry, 1989.*

Other publications
The UK pesticide guide, 1989; CAB International/British Crop Protection Council.
Weed control handbook:
Principles (Eighth edition, 1989) edited by R. A. Hance.
Recommendations (Eighth edition, 1978) edited by
J. D. Fryer and R. J. Makepeace.
Pesticides 1989. MAFF Reference Book 500. Pesticides approved under the Control of Pesticides Regulations, 1986.
Pesticides; guide to the new controls; MAFF Leaflet UL 79, revised 1988.

1

Among the free leaflets published by the Health and Safety Executive, Banyards House, 1 Chepstow Place, London W2 4TF, the following relate to herbicides:

Draft code of practice for the control of substances hazardous to health: Control of exposure to pesticide at work.

AS 18 *Storage of pesticides on farms.*

CS 19 *Storage of approved pesticides: guidance for farmers and other professional users.*

Note: Because of the rapid evolution of herbicide practice, readers should ensure that they have the most up-to-date edition of any quoted literature. It should be noted that manufacturers of herbicide products normally reprint their labels annually and may introduce changes at any reprinting.

2 Pesticide legislation

2.1 *Introduction*

The Food and Environment Protection Act 1985 Part III includes as its aims:

> "to protect the health of human beings, creatures and plants; to safeguard the environment and to secure safe, efficient and humane methods of controlling pests"

> and

> "to make information about pesticides available to the public".

More detailed conditions are laid down in the Control of Pesticides Regulations 1986 (Statutory Instrument 1510, 1986).

The Control of Pesticides Regulations 1986 affect those engaged in the application of herbicides.

2.2 *Products*

Since October 1986, *all* products marketed as pesticides must have an 'Approval' issued under the regulations. 'Approval' will be given in one of three forms.

Full approval—for an unstipulated period.

A provisional approval—for a stipulated period in order to satisfy any outstanding requirements.

An experimental permit—to enable testing and development to be carried out with a view to providing the Ministers with safety and other data.

2.3 *Conditions of approval*

Users must comply with certain conditions of approval. For most large scale uses these will be set out on the product label and may specify:

protective clothing to be worn;

crops to be treated (including minor uses, see later);

use on certain types of premises and land;

2

restrictions to use by professionals only (e.g. for glass-house fumigation);
maximum application rates;
minimum harvest intervals;
restriction of access by humans and animals to treated area;
statements about environmental protection, e.g. protection of bees.

2.4 *Minor uses*

Minor uses are defined by the MAFF as "those advantageous uses of a pesticide for which anticipated sales volume is not sufficient to persuade the manufacturer to carry out the research and development required to obtain 'full approval' for label recommendations".

As forestry frequently provides a minor use for a herbicide used more widely on other crops, this category of use is of particular interest to foresters. If approved, such minor uses are termed 'off-label approvals'.

Application to MAFF for 'off-label approvals' have been made by the Forestry Commission and other organisations engaged in forestry; but up to the time of going to press, only some applications have completed their passage through the system. Nevertheless, the text of any 'off-label approval' is available to potentially interested users. Approaches should be made to:

MAFF,
Pesticides Safety Division,
Harpenden Laboratory,
Hatching Green,
Harpenden,
Herts AL5 2BD.

Alternatively enquiries may be made to the:

Forestry Commission,
Silviculture Division,
231 Corstorphine Road,
Edinburgh, EH12 7AT

for any 'off-label approvals' thought to have been originated by the Forestry Commission. Similarly, other applicants may be approached in respect of approvals which they may have sought.

2

The approved recommendations for major uses of a herbicide will be included on the product label. A useful guide to current on-label approvals can be found in the UK Pesticide Guide (see Section 1.8).

2.5 *Operators*

Under the Control of Pesticides Regulations, general obligations are placed on all sellers, suppliers, storers and users of pesticides.

Certificates of competence are now required by those engaged in the use of pesticides. From 1 January 1989, a recognised certificate of competence is required by:

contractors applying approved pesticides unless they are working under the direct and personal supervision of a certificate holder;

those born on or after 31 December 1964 who are applying approved pesticides unless working under the direct and personal supervision of a certificate holder;

those required to supervise anyone in the above two categories, who should, but does not hold a current certificate

Pesticide operations for the purposes of certification of competence are grouped in to 'modules'. Full details are available from:

Chief Education and Training Officer,
Forestry Commission,
231 Corstorphine Road,
Edinburgh,
EH12 7AT.

Contractors who buy and sell pesticides or who advise on the use of pesticides should refer to the *Code of practice for sale or supply of pesticides* issued by MAFF for a description of their liability for:

competence and certification of any staff who may sell and/or advise on the use of pesticides;

competence and certification of staff who are responsible for storage of pesticides bought and sold.

2

Under the Control of Substances Hazardous to Health Regulations, employees are required to make an assessment. This considers the risk of exposure to pesticides proposed for use, to employees and others in the vicinity of the operation and identifies appropriate steps to be taken. The employer is then expected to take such steps if he proceeds with the pesticide application.

2.6 Use of adjuvants

The use of adjuvants is now controlled and only those appearing on lists published by MAFF may be used.

If any adjuvant appears on the list but not on the approved product label, it may be used with that product, but this would be at the user's own risk.

2.7 Aerial application of pesticides

Aerial applications are now subject to detailed rules, requiring extensive consultations. These are set out in the Control of Pesticides Regulations 1986.

2.8 Codes of practice

A series of codes of practice on pesticides are being published as part of the implementation of part III of the Food and Environment Protection Act 1985.

a. *Code of practice for agricultural and commercial horticultural use of pesticides* (statutory code: publication expected 1989).
b. *Code of practice for the use of pesticides in forestry* (non-statutory code: publication expected 1989).
c. *Code of practice for use of approved pesticides in amenity areas* (non-statutory code, first edition published February 1988).

These three codes relate to the use of pesticides. In addition there is:

Code of practice on the sale and supply including storage for supply and sale of pesticides approved for agricultural use by distributors and contractors (statutory code, publication expected 1989).

These codes of practice are intended to help users meet their obligations under the new legislation. The forestry code of practice is divided into guidance notes on the following subjects:

safe use of pesticides in forestry;

product approvals;

competence and skills of users;

protection of the operator;

protection of the environment and neighbours' crops;

safe systems for storage and handling pesticides and associated equipment including stock control and record of usage;

application equipment;

reduced volume application of pesticide from ground-based machinery.

In addition the forestry code of practice contains a working check list on the decision to use a pesticide, a working check list for operators and a check list of sources of information.

Under the Control of Substances Hazardous to Health (COSHH) Regulations, 1988 further care by employers and users of pesticides is required. A draft code on *Control of exposure to pesticides at work* has been circulated by the Health and Safety Commission. It is hoped that the various forestry, agricultural, horticultural and amenity codes will include guidance on all the requirements of the COSHH regulations as well as those under the Control of Pesticides Regulations.

2.9 *Poisons Act 1972*

Certain products are subject to the provisions of the Poisons Act 1972, the Poisons List Order 1982 and the Poisons Rules 1982 (copies of all these are obtainable from HMSO). These regulations include general and specific provisions for the labelling, storage and sale of scheduled poisons. The scheduled chemicals approved for use in the UK are listed in the UK Pesticide Guide (see Section 1.8). None of the herbicides mentioned in this Field Book are on the Poisons List.

2

2.10 *The Poisonous Substances in Agriculture Regulations 1984*

These regulations have been repealed and their provisions covered by different provisions under the Control of Substances Hazardous to Health Regulations 1988.

3 Safety precautions and safe working practices

3.1 Private sector training

Operators should check their liability under The Control of Pesticides Regulations 1986 in relation to the need for certification (see Section 2.5).

Formal training: the most effective way of achieving the level of competence necessary for certification can be arranged through the:

Forestry Training Council,
c/o Forestry Commission,
231 Corstorphine Road, Edinburgh, EH12 7AT.

Telephone: 031-334 8083.

3.2 The Forestry Safety Council and Forest Industry Safety Guides

The Forestry Safety Council (FSC) has been set up to promote all aspects of safety, particularly safe working practices throughout the forestry industry.

As an aid to maintaining the safe working standards of operators, the FSC publishes a series of Safety Guides each of which gives advice on safety in a particular forest operation.

Forest Industry Safety Guides currently available which are relevant to the application of herbicides are:

FSC 2 *Ultra low volume herbicide spraying* (rev. 4/81)
FSC 3 *Application of herbicides by knapsack spraying* (rev. 4/81)
FSC 4 *Application of granular herbicide* (rev. 4/81)
FSC 34 *First aid* (rev. 4/87)

Each Forestry Safety Guide is accompanied by a Safety Check List intended for use by supervisors, safety representatives, etc.

3

Guides are obtainable from:
The Secretary,
Forestry Safety Council,
c/o Forestry Commission,
231 Corstorphine Road,
Edinburgh, EH12 7AT.

Both operator and supervisor should be provided with a copy of the relevant leaflet, which they should read and fully understand before starting the operation.

3.3   *Routine precautions*

Users planning to apply herbicides must ensure there are proper and safe arrangements from the time of receipt of the container full of herbicide, until the final disposal of the last used container. It is necessary to observe certain general principles to achieve this objective:

provide for safe storage of herbicides;

comply with label instructions;

ensure safe transit of herbicides to work site;

wear protective clothing as specified on the label;

follow label instructions for diluting the concentrated product;

apply chemicals evenly at correct dose rate;

avoid any drift on to neighbouring crops or areas; where application is to take place near streams or lakes used as sources of drinking water liaise with the relevant water authority (see Section 3.5);

wash application equipment carefully after use to avoid contamination of crops treated subsequently;

wash and clean protective equipment regularly;

dispose of surplus chemical and empty containers safely (see Section 3.4);

keep adequate records of all operations and staff involved in the application of herbicides.

More detailed guidance on good practice can be obtained from the *Provisional code of practice for the use of pesticides in forestry* (see Section 3.6).

3.4   Safe disposal of surplus herbicides

3.4.1   Unused dilute herbicide

In well controlled application, there will be very little
unwanted dilute herbicide at the end of a day's work. Small
volumes of herbicide should be sprayed on to adjacent
ground away from water courses, avoiding susceptible crop
trees, areas specially rich in wild flowers and water-
saturated ground. The spray tank should be completely
drained.

Larger volumes of dilute herbicide, not used because of
machine failure, increased wind speed or other reasons,
should be returned overnight to a safe store and used as
soon as possible thereafter for the original intention. If,
however, manufacturers advise that dilute herbicide may
denature if kept for more than a day or so, and the herbicide cannot be used in this time, the excess liquid should
by sprayed safely on to waste ground away from water
courses as in the preceding paragraph.

3.4.2   Surplus herbicide concentrate

Unopened sound containers of herbicide, surplus to user's
requirements, should be offered back to the supplier as
soon as it is apparent that a material is surplus. Any surplus material which a supplier will not take back should be
disposed of either by prior arrangement with the local authority or by a reputable waste disposal contractor.

3.4.3   Old or deteriorated herbicide

Herbicide should not be kept beyond any date given on a
'Use Before' label or, if there is no such label, for more
than 2 years from the date of purchase. Herbicide concen-
trates showing signs of change (e.g. loss of solvent leading
to shrinkage of the container, irreversible settling out, etc.)
must not be used. Old or deteriorated herbicide should be
disposed of as for surplus herbicides (3.4.2).

3

3.5 Use of herbicides on surface water catchments that are sources of domestic water supplies

The contamination of water in general, and drinking water in particular, by herbicides is undesirable. Some catchments are especially vulnerable and the likelihood of problems can be established by discussions with the appropriate water authority. There are three principal sources of risk to surface and ground water:

spillage leading to run-off into streams;
careless disposal of waste;
careless application.

Two types of problem may be encountered. Firstly very small quantities of some herbicides (particularly phenoxy herbicides such as 2,4-D) can create severe taste and odour problems (taint) in drinking water and secondly gross pollution can occur as the result of accidental spillage.

Local water authorities in England and Wales and River Purification Boards in Scotland should be consulted if large scale use of herbicides is being considered in any surface water catchment area. In upland areas, such bodies should be able to asist in determining the extent of water catchments. It may be more difficult to determine the precise location and area of catchments of small private supplies for isolated farms and dwellings.

Consultation with water authorities or river purification boards is legally required before aerial applicatioin of pesticides on land adjacent to water. Consultation must take place not less than 72 hours before application commences (see Sections 1.6 and 2.7).

The key points when using herbicides in or near water are listed below.

● Certain herbicides are approved for use in or near water. Foresters needing to apply approved products on the banks of streams and lakes must rigorously follow the terms of the approvals for such products. Prior agreement of the water authority or river purification board is still required in such circumstances.

● Where a protective strip has been defined alongside a water course, only those products approved for use in or near water should be used on such strips.

- Herbicides not approved for use in or near water should not be applied within 10 metres of streams and water courses and within 20 metres of reservoirs. A further 10 metres should be allowed when incremental applicators are in use.
- Streams and lakes must not be used for filling or washing equipment.
- Boreholes, wells and mine shafts must not be used for disposing of waste pesticide.
- Residues must not be sprayed on to the ground within 50 metres of any borehole or well.
- In the event of any spillage threatening to enter a stream or lake, the local water authority should be informed immediately.

More detailed guidance can be found in the *Code of practice for the use of pesticides in forestry* and in the *Forests and water guidelines* (see Section 3.6).

3.6 *References for further reading*

Provisional code of practice for the use of pesticides in forestry.
Forests and water guidelines
Both publications are available from Forestry Commission Publications, Forest Research Station, Alice Holt Lodge, Wrecclesham, Farnham, Surrey GU10 4LH (Tel. 0420 22255).

The following leaflets are available free of charge from MAFF (Publications), Lion House, Willowburn Estate, Alnwick, Northumberland NE66 2PF (Tel. 0665 602881):

B 2078 *Guidelines for the use of herbicides on weeds in or near watercourses and lakes* (revised 1985)
B 2198 *Guidelines for the disposal of unwanted pesticides and containers on farms and holdings* (1985)
B 2272 *Guidelines for applying crop protection chemicals: ground crop sprayers for agriculture* (1986)
L 767   *Farm chemical stores* (1988)
L 792   *Controlled droplet application of agricultural chemicals* (1981).

Among the free leaflets published by the Health and Safety Executive, Banyards House, 1 Chepstow Place, London W2 4TF, the following relate to herbicides:

AS 6   *Crop spraying*
AS 18   *Storage of pesticides on farms*
CS 19   *Storage of approved pesticides: guidance for*
        farmers and other professional users.
Draft code of practice for the control of substances
hazardous to health: control of exposure to pesticides at
work.

The following paper may be obtained from the Water Research Centre (Environment), Medmenham Laboratory, Marlow, Bucks, SL7 2HD: Fawell, J. K. (1984). *The use of herbicides in forestry on potable water catchments.* WRC (Environment) Doc 549-M/1.

4 Grasses and grass/herbaceous broadleaved weed mixtures

4.1 *General*

Competition by grasses and herbaceous broadleaved weeds in young plantations can seriously reduce the survival and early growth of the crop trees and lead to an extended establishment period. Grasses especially can compete vigorously for light, nutrients and, in the lowlands and drier uplands, for water; effective control is therefore usually essential for successful crop establishment and growth. Because peat soils are water-retentive, moisture competition is not so severe and the need for weeding is generally less than on mineral soils.

As forest weeds, grasses can be grouped into two categories: coarse grasses which are generally tall, bulky, rank, stiff, often rhizomatous and tussocky and others which, in contrast, are known as soft grasses. Soft grasses are generally more susceptible to herbicides while coarse grasses usually show a somewhat greater resistance (see Table 1).

There are seven herbicides recommended for grass control:

ATRAZINE (various trade products)
ATRAZINE with DALAPON (Atlas Lignum)
CYANAZINE with ATRAZINE (Holtox)
GLYPHOSATE (Roundup)
HEXAZINONE (Velpar)
PROPYZAMIDE (Kerb Granules, Kerb 50 W, Kerb Flowable)
TERBUTHYLAZINE with ATRAZINE (Gardoprim A).

Their effectiveness varies according to the weed species present and the season of the year. The susceptibility of the more important grasses to these herbicides is given in Table 1.

Perennial rhizomatous grasses are the most difficult to control and require the use of residual herbicides or herbicides which are absorbed by the foliage and translocated to the rhizomes. Residual herbicides remain active in the soil and are absorbed through the roots. But they can be inactivated by the presence of organic matter in the soil. Hexazinone and dalapon, however, retain some residual activity on peat soils while atrazine and the mixtures of atrazine with cyanazine and terbuthylazine can be used as a foliage spray since they have both residual and contact action. Propyzamide has purely residual action and is not recommended for use on soils with a peat layer more than 10 cm in depth. By contrast glyphosate has foliar translocated action only and is therefore independent of soil type.

All the herbicides will have some effect on herbaceous broadleaved weeds but wherever they constitute a significant part of the weed population glyphosate (and perhaps hexazinone and terbuthylazine with atrazine) are more likely to give good results.

The following decision tree will assist in selecting a suitable herbicide according to season and weed type; see product page for specific information regarding crop tolerance soil type and other recommendations.

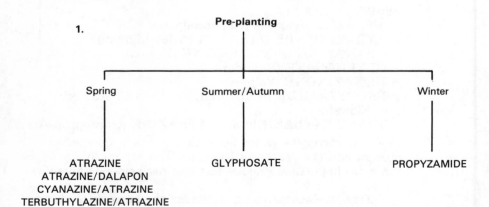

2.   **Post-planting**

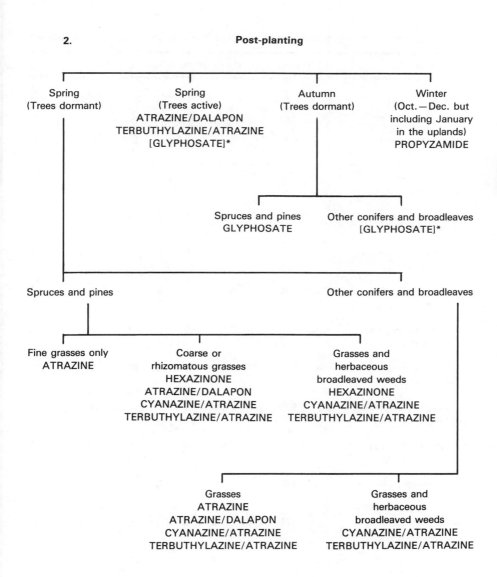

Note: Square brackets indicate that crop trees must be protected from direct contact
with the herbicide.

Table 1 Susceptibility of common grasses in the forest to recommended herbicides

Grass species		Herbicide and rate (product/ha)							
		Atrazine	Atrazine/ Dalapon	Cyanazine/ Atrazine	Glyphosate	Hexazinone	Propyzamide	Terbuthyl- azine/ Atrazine 10L	Terbuthyl- azine/ Atrazine 15L
Agropyron repens (couch grass)	(C)	MR	MS	MS	S	S	S	MR	MS
Agrostis spp. (bent grasses)		S	S	MS	S	S	S	S	S
Anthoxanthum odoratum (sweet vernal)		S	S	S	S	S	S	S	S
Arrhenatherum elatius (false oat)	(C)	MR	S	S	S	S	S	MS	S
Calamagrostis epigejos (small reed grass)	(C)	R	—	—	MS	MS	R	—	—
Dactylis glomerata (cocksfoot)	(C)	MR	MR	MS	S	S	MR	MR	S
Deschampsia caespitosa (tufted hair grass)	(C)	MR	MR	MS	S	MS	S	MS	S
Deschampsia flexuosa (wavy hair grass)		S	S	MR	S	S	S	MS	S
Festuca arundinacea (tall fescue)		MS	S	S	S	S	S	S	S
Festuca pratensis (meadow fescue)		MS	S	S	S	S	S	S	S
Festuca ovina (sheep's fescue)		S	S	S	S	S	S	S	S
Festuca rubra (red fescue)		S	S	S	S	S	S	S	S

Table 1 *continued*

Grass species	Herbicide and rate (product/ha)							
	Atrazine	Atrazine/ Dalapon	Cyanazine/ Atrazine	Glyphosate	Hexazinone	Propyzamide	Terbuthyl- azine/ Atrazine 10L	Terbuthyl- azine/ Atrazine 15L
Holcus lanatus (Yorkshire fog)	S	S	S	S	S	S	S	S
Holcus mollis (creeping soft grass)	MR	—	MS	S	S	MS	MS	MS
Molinia caerulea (C) (purple moor grass)	R	—	S	S	S	S	MR	S
Poa annua (annual meadow grass)	S	S	S	S	S	S	S	S
Poa pratensis (smooth meadow grass)	MR	S	S	S	S	S	S	S
Poa trivialis (rough meadow grass)	S	S	MS	S	S	S	S	S
Juncus spp. (rush)	R	MS	—	S	MS	R	MR	S

Abbreviations:
  S = susceptible: control should be excellent.
  MS = moderately susceptible: control should be adequate; but grass may rapidly reinvade.
  MR = moderately resistant: control may be inadequate.
  R = resistant: little effect or control obtained.
  (C) = coarse grasses (all others can be considered as soft grasses).
  — = information not available.

4

4.2 ATRAZINE

Approved products

ATRAFLOW   500 g/litre atrazine (Rhone-Poulenc
                        Environmental Products).
GESAPRIM 500 FW 500 g/litre atrazine (Ciba-Geigy).

Description

Atrazine has both foliar contact and soil action. It is most
useful on soft grasses (see Table 1) but of limited use
against herbaceous broadleaved species and the coarse
grasses (unless applied at higher rates than recommended
below).

Atrazine has no soil activity in soils with an organic peat
layer.

Planting can be carried out immediately after treatment
but the level of weed control will be reduced if the soil is
badly disturbed.

Crop tolerance

Conifers: all the major forest species are tolerant to overall
application except NS, WH and EL which are sensitive during the growing season and should only be treated before
the start of bud burst.

Broadleaves: all are sensitive while in leaf and should
only be treated before the start of bud burst in the spring.

Product rate

Apply 10 litres of product per treated hectare.

Methods of application

*Pre-plant (overall) or post-plant (directed, overall, band
or spot application)*
Tractor mounted equipment
Boom sprayer at MV.
Ulvaforest low speed rotary atomiser at VLV.
Handheld equipment
Knapsack sprayer at MV.
Knapsack sprayer with 'VLV' nozzle at LV.
Herbi low speed rotary atomiser at VLV.

Forestry Spot Gun at LV.

Refer to Section 11 for details of applicator and correct calibration.

Timing of application

February to May. February and March applications usually give the best results. Applications in May and June will give effective weed control but also slight crop damage (and severe damage to the more sensitive species NS, WH and larch: see above).

4

Additional information

1. Weed control

a. Atrazine has high solubility, it is therefore unsuitable for use on very light soils, soils with poor structure, i.e. man-made sites, recently constructed sites (which are particularly prone to frost or drought cracking) or sites prone to waterlogging.

Crops on light calcareous or sandy soils seem to suffer relatively more through damage from atrazine than crops on heavier textured soils.

b. Do not apply atrazine to unhealthy or badly planted trees.

c. a and b above are particularly important for broadleaved species.

d. The risk to broadleaved trees can be reduced by the use of directed application.

e. Care must be taken when atrazine is used on steep slopes; heavy rain soon after application can cause creep and surface run-off before the chemical has penetrated the soil.

f. Atrazine does not give an acceptable level of weed control on soils with an organic peat layer.

g. *Holcus mollis*, although a soft grass, is not readily controlled.

h. Do not use atrazine on NS intended for Christmas trees.

2. Protective clothing

Read product label for protective clothing and equipment requirements and check there are no items required in addition to the Forestry Commission recommendations in Section 10.

3. Special precautions

Atrazine concentrate can be harmful to fish; do not contaminate ponds, waterways and ditches with chemical or used containers.

The label on the herbicide container has been designed for your protection—**ALWAYS READ THE INSTRUCTIONS ON THE LABEL**.

4

4.3 *ATRAZINE WITH DALAPON*

Approved product
ATLAS LIGNUM 100 g/kg atrazine+100 g/kg dalapon
(Atlas Interlates Ltd.).

Description
A mixture of atrazine and dalapon in granular form. The
mode of action is similar to that of atrazine alone (see
Section 4.2) but the inclusion of dalapon extends its
activity to coarse grasses and on to peat soils. Late
summer applications may control some herbaceous broadleaved weeds. Planting can be carried out immediately after
treatment.

Crop tolerance
SS, SP, LP, GF, NF and RC are tolerant. NS, CS, DF,
larches, WH and broadleaves are rather more sensitive.
Application rates for these species should not exceed 30 kg
of the product per treated hectare.

Product rate
Apply 40 kg of product per treated hectare, or 30 kg if
sensitive crop species are present (see Crop tolerance
above).

Method of application
*Pre-plant or post-plant (overall, band or spot
application)*
Handheld equipment
Pepperpot applicator.
Moderne applicator.
   Refer to Section 11 for details of applicators and correct
calibration.

Timing of application
Mid March to May prior to bud burst is the optimum
period. Earlier applications may be effective if there is
active grass growth. June to August applications will be
less effective and the dalapon component will cause damage
if it sticks to the tree foliage.

4

4

Additional information

1. Weed control

a. Even application of granules is essential to avoid crop damage.

b. Moisture is necessary to activate Atlas Lignum granules (to release the herbicide from the inert granule and wash it into the soil); reduced weed control may result if prolonged dry conditions follow application.

c. Where a thick layer of old plant debris is present at the time of application reduced weed control may result because the herbicide is prevented from reaching the roots of weeds.

d. Care must be taken when application is made in windy conditions otherwise the granules will not land in the target area.

e. If Atlas Lignum granules are applied when the trees are in leaf, application must take place when the tree foliage is dry to prevent the fine granules sticking to the foliage and causing damage.

f. *Calamagrostis epigejos* (small reed grass) is likely to be resistant.

2. Protective clothing

Read product label for protective clothing and equipment requirements and check there are no items required in addition to the Forestry Commission recommendations in Section 10.

3. Special precautions

Atlas Lignum concentrate can be harmful to fish; do not contaminate ponds, waterways and ditches with chemical or used containers.

The label on the herbicide container has been designed for your protection—**ALWAYS READ THE INSTRUCTIONS ON THE LABEL.**

4.4 CYANAZINE WITH ATRAZINE

Approved product
HOLTOX 250 g/litre cyanazine 250 g/litre atrazine (Rhone-Poulenc Environmental Products).

Description
Cyanazine/atrazine has both foliar and soil action and will control a wide range of grasses and annual herbaceous broadleaved weeds. Effective weed control can be achieved when applied either to bare soil or established weeds. When Holtox is applied to dormant weeds the effect will take considerably longer to appear than if it is applied to actively growing weeds.

4

The level of control obtained from the residual action of this herbicide is reduced where an organic peat layer is present.

Planting can be carried out immediately after treatment but the level of weed control will be reduced if the soil is badly disturbed following application.

Crop tolerance
Conifers: SS, NS, SP, CP, LP, larches, GF, DF and OMS are tolerant of treatment before the start of bud burst; all species are sensitive during the growing season.

Broadleaves: ash, alder, birch, beech, oak, rowan and whitebeam are tolerant before the start of bud burst; refer to manufacturer for tolerance of other species.

Product rate
Apply 10–13.5 litres of product per treated hectare. Use the 10 litre rate when application is made to bare ground, immature weeds or light soils.

Methods of application
Pre-plant (overall) or post-plant (directed, overall, band or spot application)
Tractor mounted equipment
Boom sprayer at LV or MV.
Ulvaforest low speed rotary atomiser at VLV.

Handheld equipment
Knapsack sprayer at MV.
Knapsack sprayer with 'VLV' nozzle at LV.
Herbi low speed rotary atomiser at VLV.
Forestry Spot Gun at LV.
   Refer to Section 11 for details of applicator and correct calibration.

4

Timing of application
Optimum weed control is obtained by applications made just as weed growth commences in the spring (see Crop tolerance above).

Additional information
1. Weed control
a. For optimum weed control Holtox should be applied to moist soil and some rain should follow application; reduced weed control or a delay in weed control may result if prolonged dry conditions follow application.
b. Do not apply Holtox to unhealthy or badly planted trees.
c. Care must be taken when Holtox is used on steep slopes; heavy rain soon after application can cause creep and surface run-off before the chemical has penetrated the soil.
d. On sites with an organic peat layer the residual properties of Holtox will be reduced. On these sites control will be achieved through the contact action of this herbicide; it must therefore be applied when the weeds are actively growing (see Crop tolerance).
e. Established *Holcus mollis*, although a soft grass, may rapidly reinvade treated areas.

2. Protective clothing
Read product label for protective clothing and equipment requirements and check there are no items required in addition to the Forestry Commission recommendations in Section 10.

3. Special precautions

Holtox concentrate can be harmful to fish; do not contaminate ponds, waterways and ditches with chemical or used containers.

The label on the herbicide container has been designed for your protection—**ALWAYS READ THE INSTRUCTIONS ON THE LABEL.**

4

4

4.5 GLYPHOSATE

Approved product
ROUNDUP 360 g/litre glyphosate (Monsanto).

Description
A translocated herbicide taken up by the foliage and conveyed to the roots. It causes chlorosis and eventual death of leaves and then kills roots and shoots.

Roundup controls a wide range of weeds including grasses, herbaceous broadleaved weeds, bracken, heather and woody weeds. When applied late in the growing season, the main effect is obtained in the following year.

On contact with the soil Roundup is inactivated and quickly broken down. Planting can be carried out immediately after treatment, but a minimum of 5 days should be allowed before cultivation and the breaking up of rhizomes and roots.

Crop tolerance
SS, NS, SP, CP, LP, RC and LC: will tolerate overall sprays provided leader growth has hardened. Hardening can occur as early as the end of July or may be delayed until the end of October in some locations and seasons. To avoid damage to lammas growth, herbicide sprays must be directed away from leaders. During the active growing season trees must be guarded or the spray directed to avoid contact with the crop.

DF and NF: as above but much more sensitive.

Broadleaved trees, larch and other conifers will not tolerate overall applications: always use a guard, a weedwiper or a directed spray to avoid contact with the foliage and immature bark of crop trees.

Product rate
1. Sprays for weed control pre and post-planting

Upland Britain: 2.0 litres ⎫
⎬ of product per treated hectare diluted in water to a total volume appropriate for
Lowland Britain: 1.5 litres ⎭ the applicator and method selected.

2. Direct application
1 part of product diluted in 6 parts of water through a direct applicator. On wetter upland sites use 1 part of product diluted in 3 parts of water.

Where weeds such as bracken and bramble are to be controlled in addition to grass and broadleaved weeds then the rate must be increased to:
2 litres per hectare for the control of bracken (see Section 5);
3 litres per hectare for the control of bramble (see Section 7).

4

Methods of application
1. Pre-plant (overall or band application)
Note: a preplant band application is only feasible when the strip to be treated is identified by rip cultivation.

Tractor mounted equipment
Boom sprayer at LV or MV.
Ulvaforest low speed rotary atomiser at VLV.

Handheld equipment
Knapsack sprayer at MV.
Knapsack sprayer with 'VLV' nozzle at LV.
Herbi low speed rotary atomiser at VLV.

2. Post-plant (directed, overall, band or spot application)
Tractor mounted equipment
Boom sprayer at LV or MV.
Ulvaforest low speed rotary atomiser at VLV.

Handheld equipment
Knapsack sprayer at MV.
Knapsack sprayer with 'VLV' nozzle at LV.
Herbi low speed rotary atomiser at VLV.
Forestry Spot Gun for spot application at LV.
Weedwiper for direct application; uncalibrated (see Product rate).

Refer to Section 11 for details of applicator and correct calibration.

Timing of application

Roundup can be applied at any time of year when vegetation is actively growing but is most effective on broadleaved weeds when they are close to flowering but before
senescence.

When crop trees are present this will influence the
timing and/or method of application (see Crop tolerance).

Additional information

1. Weed control

a. Roundup applied later than June will be too late to
   lessen the effect of weed competition in the current
   season.
b. Roundup is most effective on moist vegetation and when
   relative humidity is high and the air is warm (e.g.
   $15°C+$).
c. Reduced weed control may result when weeds are under
   stress, e.g. frost or drought.
d. Direct application using the weedwiper, achieves
   maximum control when the vegetation is actively
   growing and under 0.3 m in height. In taller vegetation
   and where a large number of seed heads are present the
   degree of control will be reduced. Care must be taken to
   avoid low tree branches amongst weeds.
e. Heavy rainfall within 24 hours of application may reduce
   the herbicide's effectiveness by preventing sufficient
   foliar absorption. The addition of Mixture B at 2% of
   final spray volume will improve reliability in these
   circumstances. The addition of Mixture B will reduce
   crop tolerance and it must only be used pre-plant or
   post-plant as a directed spray.
f. Diluted Roundup may denature after 2 to 3 days. Where
   possible use tap water as the diluent and only mix sufficient for the day's programme.

2. Protective clothing

Read product label for protective clothing and equipment
requirements and check there are no items required in
addition to the Forestry Commission recommendations in
Section 10.

3. Special precautions

Roundup concentrate can be harmful to fish; do not contaminate ponds, waterways and ditches with chemical or used containers.

The label on the herbicide container has been designed for your protection—**ALWAYS READ THE INSTRUCTIONS ON THE LABEL.**

4

4.6 *HEXAZINONE*

Approved product
VELPAR LIQUID 240 g/litre hexazinone (Selectokil).

Description
A residual translocated herbicide which has some foliar contact action but is primarily taken in through the roots. Velpar controls a wide range of grasses and many herbaceous broadleaved weeds, and is effective on organic soils.

When using Velpar, trees must have been established for at least 1 year.

Crop tolerance
SS, NS, SP, CP and LP: will tolerate overall sprays provided leader growth has hardened. Hardening can occur as early as the end of July or may be delayed until the end of October in some locations and seasons. During the active growing season trees must be guarded or the spray directed to avoid contact with the crop.

All other species are susceptible to Velpar and must not be treated. Avoid spray drift on to deciduous trees including larch.

Product rate
Apply 7 litres of product per treated hectare in water.

Method of application
Pre-plant (overall application) or post-plant (directed, overall, band or spot application)
Handheld equipment
Knapsack sprayer (with a guard if required) at MV (not less than 240 l/ha). A large diameter flood jet nozzle and low pressure should be used to avoid the production of fine droplets.
Forestry Spot Gun.

Refer to Section 11 for details of applicator and correct calibration.

Timing of application
Velpar is effective if applied from mid-February to August but best results are achieved by April–May applications, to small actively-growing weeds. The presence of a crop limits overall application to early spring, before the start of bud burst, and the period after shoot growth has ceased (see Crop tolerance above).

Additional information

1. Weed control
a. Moisture is necessary to activate Velpar; reduced weed control may result if prolonged dry conditions follow application.
b. Velpar should not be applied to unhealthy or badly planted trees.

2. Protective clothing
Read product label for protective clothing and equipment requirements and check there are no items required in addition to the Forestry Commission recommendations in Section 10.

3. Special precautions
a. Velpar is an eye irritant: avoid all contact with eyes.
b. Velpar should not be applied via controlled droplet applicators.
c. Velpar concentrate can be harmful to fish; do not contaminate ponds, waterways and ditches with chemical or used containers.

   The label on the herbicide container has been designed for your protection—**ALWAYS READ THE INSTRUCTIONS ON THE LABEL.**

4.7 *PROPYZAMIDE*

Approved products
KERB GRANULES 40 g/kg propyzamide (PBI).
KERB 50W 500 g/kg propyzamide: wettable powder (PBI).
KERB FLOWABLE 400 g/litre propyzamide (PBI).

Description
A soil acting herbicide which slowly volatilises in cold soil and is taken up by germinating weeds and through the roots of existing weeds, especially grasses. A wide range of herbaceous broadleaved weeds are susceptible from germination to the two true leaf stage. But herbaceous broadleaved weeds which emerge late in the season will only be partially controlled.

Propyzamide slowly breaks down in the soil, lasting for 3—6 months.

Crop tolerance
All commonly planted forest tree species are tolerant.

KERB GRANULES
Product rate
Apply at 38.0 kg of granules per treated hectare.

Method of application
Pre or post-plant (overall, band or spot application)
Apply through the Pepperpot or Moderne applicator.

KERB 50W
Product rate
Apply as 3 kg of product per treated hectare in water.

Method of application
Pre or post-plant (overall, band or spot application)
Tractor mounted equipment
Boom sprayer at MV.

Handheld equipment
Knapsack sprayer at MV.
Knapsack sprayer with 'VLV' nozzle at LV.
Forestry Spot Gun for spot application at LV.
For best results mix the powder into a smooth paste with a
small amount of water, slowly add more water, stirring all
the time, until the mix is sloppy, then add the remaining
water up to full dilution.

4

KERB FLOWABLE

Product rate
Apply at 3.75 litres of product per treated hectare.

Method of application
Pre or post-plant (overall, band or spot application)
Tractor mounted equipment
Boom sprayer at MV.
Ulvaforest low speed rotary atomiser at VLV.

Handheld equipment
Knapsack sprayer at MV.
Knapsack sprayer with VLV nozzle at LV.
Herbi low speed rotary atomiser at VLV.
Forestry Spot Gun at LV.
    When applying Kerb Flowable via the Herbi or
ULVAforest a non-ionic wetting agent (such as PBI
Spreader or Agral) should be added at the rate of 0.5% of
final spray volume.
    Refer to Section 11 for details of applicator and correct
calibration.

Timing of application
Apply from 1 October to 31 January, north of a line from
Aberystwyth to London.
    Apply from 1 October to 31 December, south of a line
from Aberystwyth to London or on peat or peaty gley soils
(see Additional information 1b).

Additional information

1. Weed control

a. Although propyzamide can be used very effectively for pre-planting applications, the effect of the herbicide applied in winter cannot often be seen by the time of normal planting.

b. Organic soils decrease the activity of propyzamide and treatment of soils with a depth of peat greater than 10 cm is not recommended.

c. The following grasses show some resistance to propyzamide:
Dactylis glomerata
Holcus mollis
Calamagrostis epigejos.

d. Planting can be carried out immediately pre- or post-treatment.

2. Protective clothing

Read product label for protective clothing and equipment requirements and check there are no items required in addition to the Forestry Commission recommendations in Section 10.

3. Special precautions

Propyzamide concentrate can be harmful to fish; do not contaminate ponds, waterways and ditches with chemical or used containers.

The label on the herbicide container has been designed for your protection—**ALWAYS READ THE INSTRUCTIONS ON THE LABEL.**

4.8 *TERBUTHYLAZINE WITH ATRAZINE*

Approved products
GARDOPRIM A 500FW 400 g/litre terbuthylazine
100 g/litre atrazine (Ciba-Geigy).

Description
Terbuthylazine/atrazine has both foliar contact action and soil action. Most annual herbaceous broadleaved weeds and grasses are controlled at the lower rate, while perennial and well established weeds must be treated with the higher rate to achieve effective control.

The level of control obtained from the residual action of this herbicide is reduced where an organic peat layer is present.

Planting can be carried out immediately after treatment but the level of weed control will be reduced if the soil is badly disturbed.

Crop tolerance
Conifers: SS and DF are tolerant of overall applications at any time of year. NS, SP, CP, LP, RAP, NF, EL, JL should only be treated before the start of bud burst.

Broadleaves: ash, beech, oak and sycamore are tolerant of overall applications when dormant and should only be treated before the start of bud burst.

All other species are sensitive and should not be treated.

Product rate
Apply 10–15 litres of product per treated hectare depending on the weed species to be controlled (see Table 1).

Methods of application
Pre-plant (overall) or post-plant (directed, overall, band or spot application)
Tractor mounted equipment
Boom sprayer at MV.
Ulvaforest low speed rotary atomiser at VLV.
Handheld equipment
Knapsack sprayer at MV.
Knapsack sprayer with 'VLV' nozzle at LV.

4

Forestry Spot Gun at LV.
    Refer to Section 11 for details of applicator and correct calibration.

Timing of application
Optimum weed control is obtained by applications made just as weed growth commences in the spring (see Crop tolerance above).

Additional information

1. Weed control
a. Gardoprim A is very insoluble, moisture is therefore necessary to activate this herbicide, reduced weed control or a delay in weed control may result if prolonged dry conditions follow application.
b. Do not apply Gardoprim A to unhealthy or badly planted trees.
c. Care must be taken when Gardoprim A is used on steep slopes; heavy rain soon after application can cause creep and surface run-off before the chemical has penetrated the soil.
d. On sites with an organic peat layer the residual properties of Gardoprim A will be reduced. On these sites control will be achieved through the contact action of this herbicide; it must therefore be applied when the weeds are actively growing (see Crop tolerance).
e. Gardoprim A can be applied in wet weather as long as conditions are not likely to lead to surface run-off from the site. When contact action is most important (see d above) drier weather is desirable to prevent the herbicide being washed off the foliage of the weeds.
f. Established *Holcus mollis*, although a soft grass, may rapidly reinvade treated areas.
g. Gardoprim A is not recommended for use on Christmas trees.
h. Gardoprim A is not approved for application via handheld controlled droplet applicators.

2. Protective clothing
Read product label for protective clothing and equipment requirements and check there are no items required in addition to the Forestry Commission recommendations in Section 10.

3. Special precautions

Gardoprim A concentrate can be harmful to fish; do not contaminate ponds, waterways and ditches with chemical or used containers.

The label on the herbicide container has been designed for your protection—**ALWAYS READ THE INSTRUCTIONS ON THE LABEL.**

4

5 Bracken

5.1 *General*

Bracken competes strongly with young trees for light during the later part of the growing season. At the end of the year it collapses and can smother and flatten small trees with its weight, increasingly so if snow lies on top of them both. Bracken is rarely sufficiently advanced in spring to afford protection from frosts and is not worth retaining for this purpose. Weeds which take over from bracken can usually be easily controlled by chemicals.

Ploughing does give some control of bracken for the first season but on sites where bracken is vigorous, the stems on either side of the plough ridge will overgrow conifers.

If a crop is present, chemical control must be followed at least one month later by hand cutting before the fronds collapse on to the trees and cause damage. Whenever possible, herbicide should be applied pre-planting to avoid this problem.

The herbicides used for bracken control are:

    ASULOX   (asulam);

    ROUNDUP (glyphosate).

Both these products are translocated from the fronds to the rhizomes where they have a herbicidal effect. The result is that frond growth the following year is prevented or retarded.

Asulox gives slightly better control of bracken but no other weeds are controlled. Conifer tolerance is high.

Roundup gives adequate control but can cause damage to the crop particularly if it is used in mid-summer. If bracken is mixed with other weeds (e.g. brambles) then Roundup should be chosen for its wide spectrum of weed control.

5.2   ASULAM

Approved product
ASULOX 400 g/litre asulam (Embetec Crop Protection).

Description
Asulam is a post-emergence, translocated herbicide which is
taken up by the foliage and translocated to the rhizomes.
Herbicidal symptoms are virtually absent in the year of
spraying but growth of the bracken is then retarded or fails
entirely the following season. Control may last from 1 to 4
years, or more depending on the rate applied and the date
of application. At least 6 weeks should elapse between
applying Asulox and subsequent planting.

5

Crop tolerance
All the major forest species of conifers are tolerant except
WH. Mature trees of WH are tolerant of recommended
rates but young trees of WH may show slight chlorosis and
check at the highest rates and earliest dates of application.
   But WH will tolerate up to 7 litres of product per
hectare during August and early September.
   Beech, birch, elm and poplar: as for conifers.
   Other broadleaves: susceptible. Trees should be protected
by a guard or the spray directed to avoid the crop.

Product rate
Apply 5 litres (early season), 10 litres (late season), of
product per hectare diluted in water.

Methods of application
*Pre-plant (overall application) or post-plant (directed or
overall)*
Tractor mounted equipment
Boom sprayer at MV.
Ulvaforest low speed rotary atomiser at VLV.
Handheld equipment
Knapsack sprayer at MV.
Knapsack sprayer with 'VLV' nozzle at LV.
Mistblower at LV.
ULVA high speed rotary atomiser at VLV.

Aerial (check requirements of Control of Pesticides Regulations). Apply Asulox in 55 litres of water per hectare using either Delavan 'Raindrop' or conventional nozzles.

Refer to Section 11 for details of applicator and correct calibration.

Timing of application

Early July to late August (early August in northern Britain). Best results are obtained by application just as the frond tips have unfurled and formed an almost complete canopy. Treatment at this stage may reduce the need to hand-cut the dead stems at the end of the growing season. The later in the season application is made the higher is the rate of Asulox required to obtain control.

Additional information

1. Weed control

a. To obtain optimum control of bracken a minimum of one month should elapse after treatment before cutting the bracken down or ploughing the ground.
b. Access for spraying should be made by pushing fronds aside and not by cutting.
c. Heavy rainfall within 24 hours of application may reduce the herbicide's effectiveness by preventing sufficient foliar absorbtion.
d. Reduced weed control may result when weeds are under stress e.g. frost or drought or when application is made during conditions of high temperature and low humidity.
e. For best results spray before the onset of senescence.

2. Protective clothing

Read product label for protective clothing and equipment requirements and check there are no items required in addition to the Forestry Commission recommendations in Section 10.

3. Special precautions

Asulox concentrate can be harmful to fish; do not contaminate ponds, waterways and ditches with chemical or used containers.

The label on the herbicide container has been designed for your protection—ALWAYS READ THE INSTRUCTIONS ON THE LABEL.

5.3 GLYPHOSATE

Approved product
ROUNDUP 360 g/litre glyphosate (Monsanto).

Description
A translocated herbicide taken up by the foliage and conveyed to the rhizomes. It causes chlorosis and eventual death of fronds and prevents regrowth.

Roundup controls a wide range of weeds including grasses, herbaceous broadleaved weeds, bracken, heather and woody weeds. It is particularly effective for bracken in mixture with these other weed types. On bracken some dieback of foliage can be expected in the year of application. In the following season rhizomes fail to send out fronds.

On contact with the soil Roundup is inactivated and quickly broken down. Planting can be carried out immediately after treatment, but a minimum of 5 days should be allowed before cultivation and the breaking up of rhizomes and roots.

Crop tolerance
SS, NS, SP, CP, LP, RC and LC: will tolerate overall sprays provided leader growth has hardened. Hardening can occur as early as the end of July or may be delayed until the end of October in some locations and seasons. To avoid damage to lammas growth, herbicide sprays must be directed away from leaders. During the active growing season trees must be guarded or the spray placed to avoid contact with the crop.

DF and NF: as above but much more sensitive.

Broadleaved trees, larch and other conifers will not tolerate overall applications: always use a guard, or a directed spray to avoid contact with the foliage and immature bark of crop trees.

Product rate
Apply 2.0 litres of product per hectare.

Where bramble is to be controlled in addition to bracken then the rate must be increased to 3.0 litres per hectare (see Section 7).

Methods of application
Pre-plant (overall application) or post-plant (overall or directed application)
Tractor mounted equipment
Boom sprayer at LV or MV.
Ulvaforest low speed rotary atomiser at VLV.

Handheld equipment
Knapsack sprayer at MV.
Knapsack sprayer with 'VLV' nozzle at LV.
Mistblower at LV.
ULVA high speed rotary atomiser at VLV. For full effect, dilute the herbicide with at least 5 times the volume of water.

Refer to Section 11 for details of applicator and correct calibration.

Timing of application
Best results will be obtained from applications made during July and August; after frond tips have uncurled but before senescence.

Additional information
1. Weed control
a. Roundup applied to control bracken in July and August will be too late to lessen the effect of weed competition or smothering in the current season.
b. Roundup is most effective on moist vegetation and when relative humidity is high and the air is warm (e.g. 15°C+).
c. Reduced weed control may result when weeds are under stress, e.g. drought.
d. Heavy rainfall within 24 hours of application may reduce the herbicide's effectiveness by preventing sufficient foliar absorption. The addition of Mixture B at 2% of final spray volume will improve reliability in these circumstances. The addition of Mixture B will reduce crop tolerance and it must only be used pre-plant or post-plant as a directed spray.
e. Diluted Roundup may denature after 2 to 3 days. Where possible use tap water as the diluent and only mix sufficient for the day's programme.

f. Access for spraying should be made by pushing the fronds aside and not by cutting.
g. A minimum of one month should elapse after treatment before cutting the bracken down or ploughing the ground.
h. On bracken mixed with woody weeds, mistblowing is usually the most effective of the handheld methods but it is also likely to inflict a higher degree of damage on unhardened crop trees.

2. Protective clothing
Read product label for protective clothing and equipment requirements and check there are no items required in addition to the Forestry Commission recommendations in Section 10.

3. Special precautions
Roundup concentrate can be harmful to fish; do not contaminate ponds, waterways and ditches with chemical or used containers.

The label on the herbicide container has been designed for your protection—ALWAYS READ THE INSTRUCTIONS ON THE LABEL.

5

6 Heather

6.1 General

On sites where the availability of mineral nitrogen limits tree growth, and where the dominant vegetation is heather (*Calluna vulgaris*), nitrogen deficiency may develop in certain species of conifer. This may need to be alleviated by complete spraying to kill the heather.

Sitka spruce is by far the most important species in this context and unless otherwise stated all the recommendations given here relate to it. NS, WH, GF, NF, DF and CP may also be severely checked by heather competition.

The need to control heather can often be avoided by:
    planting a non-susceptible species;
    burning the heather before ploughing;
    restocking felled areas before the heather has time to invade;
    planting spruce in mixture with SP, larches or LP,
Herbicides recommended for heather control are:
    2,4-D ESTER (Silvapron D for ULV application via a rotary atomiser and emulsifiable concentrate for MV application);
    ROUNDUP (glyphosate).

2,4-D ester is the generally recommended herbicide. It can be used selectively during the growing season although at the rates required to kill heather some damage is to be expected to the crop.

Roundup can be used in late season when trees are dormant and this may provide the best means of controlling heather amongst trees less than 1 metre in height. Roundup does not taint water and has a very low mammalian toxicity and may therefore be the best choice for water catchment areas. Roundup is however likely to cause some damage to crop trees at the rates required to give a good heather kill.

The level of heather control obtained is often variable depending on the vigour of the heather involved. On sites with a high nitrogen status, a rapid turnover of nutrients and a large leaf area, heather is usually easy to kill.

6.2 *2,4-D ESTER*

Approved products

ULV formulation
SILVAPRON D   400 g/litre   (BP).

Emulsifiable concentrate of a low volatile ester
BH 2,4-D ESTER 50   465 g/litre   (Rhone-Poulenc).

Description

2,4-D is a plant growth regulating herbicide to which many herbaceous and woody broadleaved species are susceptible. It is absorbed mainly through aerial parts of the plant but also through the roots. Because of the oil-based formulation, applications of 2,4-D ester are relatively rainfast. The hormonal activity browns the heather shortly after treatment but exhibits its full effect in the following season.

There is a risk to bees through ingestion when spraying heather in flower. This risk can be minimised by good liaison with bee-keepers.

One month should elapse between treatment and subsequent planting.

Crop tolerance

SS, NS and OMS are moderately tolerant of overall sprays provided leader growth has hardened. Hardening can occur as early as the end of July or may be delayed until the end of October in some locations and seasons.

SP, CP, DF, RC, GF and NF are rather less tolerant and spray should be directed away from foliage, particularly leading shoots.

LP, WH and larches are sensitive.

Broadleaves are very sensitive to 2,4-D.

Hot weather at the time of spraying may increase crop damage.

6

ULTRA LOW VOLUME FORMULATION
Product rate

Area	Soil type	Mid-July to mid-August	Second half of August
Northern Britain	Peat	10 litres/ha	12 litres/ha
	Mineral	15 litres/ha	15 litres/ha
Southern Britain	All sites	10 litres/ha	12 litres/ha

Method of application
Pre-plant (overall application) or post-plant (overall or directed application); see Crop tolerance
Apply by high speed rotary atomiser without dilution through the ULVA using the incremental method of application (see Section 11.5.5) or the tractor mounted Ulvaforest.

Post-plant directed application
The ULVA can be used to apply a directed inter-row spray by inverting the applicator and holding the head at about 45 degrees to the ground.

The mean tree height should be at least 1 metre to prevent contact between the spray and the leading shoots of the trees.

As an alternative, the Herbi can be used but the level of heather control achieved is lower than with ULV equipment. On the other hand crop tolerance is better because the spray can be more easily directed. Inter-row spraying with the Herbi can therefore begin as early as May or June (at the same rates as for mid-July), increasing the 10 and 12 litre rates in the above table by 25% will achieve an effect equivalent to that of the ULVA.

EMULSIFIABLE CONCENTRATE
Product rate

Soil type	May	June to mid-August	Second half of August
Peat	10 litres	8 litres	10 litres
Mineral	12 litres	10 litres	12 litres

Method of application
Post-plant (directed application)
The appropriate volume of emulsifiable concentrate is
applied in water through the following applicators:
    knapsack sprayer at MV;
    knapsack sprayer with 'VLV' nozzle at LV.

Timing of application
See tables in preceding paragraphs but note that
application earlier than mid-July should only be applied as
a directed spray so as to avoid serious crop damage.

**Additional information
(applicable to both ULV formulation and emulsifiable
concentrate)**

6

1. Weed control
a. For optimum control of heather apply 2,4-D when the
   heather is in flower, but because of the risk to bees
   following application bee-keepers in the locality must be
   given adequate notice.
b. It is important to get good coverage of heather foliage
   and stems for satisfactory control.
c. Broadleaved plants are very sensitive to 2,4-D especially
   in warm weather; avoid spraying when wind would cause
   drift and damage to neighbouring crops or trees.
d. To reduce the risk of crop damage trees should be 1
   metre tall to minimise contact of spray mixture and
   leading shots.
e. Hot weather at the time of application may increase the
   risk of crop damage.
f. Heather should be dry at the time of application; heavy
   rain soon after application may reduce the level of
   control achieved.

2. Protective clothing
Read product label for protective clothing and equipment
requirements and check there are no items required in
addition to the Forestry Commission recommendations in
Section 10.

3. Special precautions
a. Special precautions are required in water catchment areas to avoid water taint—see Section 3.5.
b. If possible 2,4-D should not be applied in areas visited regularly by the public in significant numbers, and where edible fruits or plants are likely to be exposed to spray. If, however, spraying does take place in such areas appropriate warning notices must be erected. Such signs should remain in position as long as treated fruit appear wholesome.
c. Heather flowering will probably be affected by application of 2,4-D earlier than August. Beekeepers should be advised not to site hives on the areas to be sprayed.
d. Harmful to fish; do not contaminate ponds, waterways or ditches with chemical or used container.

   The label on the herbicide container has been designed for your protection—**ALWAYS READ THE INSTRUCTIONS ON THE LABEL.**

6

6.3   *GLYPHOSATE*

Approved product
ROUNDUP 360 g/litre glyphosate (Monsanto).

Description
A translocated herbicide taken up by the foliage and conveyed to the roots. It causes chlorosis and eventual death of leaves and then kills roots and shoots.

Roundup controls a wide range of weeds including grasses, herbaceous broadleaved weeds, bracken, heather and woody weeds. The herbicidal effects take some time to develop on heather and the full response is not evident until the following growing season.

On contact with the soil Roundup is inactivated and quickly broken down. Planting can be carried out immediately after treatment, but a minimum of 5 days should be allowed before cultivation and the breaking up of rhizomes and roots.

Crop tolerance
At the rates used for heather control spruces and pines are only moderately tolerant to overall sprays of Roundup after new growth has hardened. For this reason only directed sprays are recommended post planting.

Hardening can occur as early as the end of July or may be delayed until the end of October in some locations and seasons. The spray should be directed to avoid leaders especially in seasons when lammas growth occurs. Other species should not be sprayed when using the rates applicable for heather control.

Product rates
On mineral soils: 6 litres of product per hectare.
On peaty soils: 4 litres of product per hectare.

Methods of application
Pre-plant (overall application)
Tractor mounted equipment
Boom sprayer at LV or MV.
Ulvaforest low speed rotary atomiser at VLV.

Handheld equipment
Knapsack sprayer at MV.
Knapsack sprayer with 'VLV' nozzle at LV.
Herbi low speed rotary atomiser at VLV.

Post-plant (directed application)
Handheld equipment
Knapsack sprayer at MV.
Knapsack sprayer with 'VLV' nozzle at LV.
Herbi low speed rotary atomiser at VLV.
　　Refer to Section 11 for details of applicator and correct calibration.

Timing of application
Late August to end of September after new growth on crop trees has hardened (see Crop tolerance). This is also the optimum time for the control of heather pre-planting.

Additional information
1. Weed control
a. Roundup applied in August and September to control heather will be too late to lessen the effect of weed competition in the current season.
b. Roundup is most effective on moist vegetation and when relative humidity is high and the air is warm (e.g. 15°C+).
c. Reduced weed control may result when weeds are under stress, e.g. drought.
d. Heavy rainfall within 24 hours of application may reduce the herbicide's effectiveness by preventing sufficient foliar absorption. The addition of Mixture B at 2% of final spray volume will improve reliability in these circumstances. The addition of Mixture B will reduce crop tolerance and it must only be used pre-plant or post-plant as a directed spray.
e. Diluted Roundup may denature after 2 to 3 days. Where possible use tap water as the diluent and only mix sufficient for the day's programme.

2. Protective clothing
Read product label for protective clothing and equipment requirements and check there are no items required in

addition to the Forestry Commission recommendations in Section 10.

3. *Special precautions*

a. Roundup concentrate can be harmful to fish; do not contaminate ponds, waterways and ditches with chemical or used containers.

b. While bees are unaffected by glyphosate, heather flowering will probably be affected by applications earlier than August. Beekeepers should be advised not to site hives on areas to be sprayed.

The label on the herbicide container has been designed for your protection—**ALWAYS READ THE INSTRUCTIONS ON THE LABEL.**

6

7 Woody weeds

7.1 General

This group of weeds contains a wide range of species
including brambles, climbers, shrubs and all types of tree.
Species such as birch in some circumstances are part of the
crop and in others are unwanted and threaten the crop.
Such situations require a precise definition of management
objectives and constraints (such as the present and future
amenity effects of broadleaved components of a stand).

The biological similarity of woody weeds to crop trees
can make selective chemical control difficult or impossible.
As perennial woody plants, many with the ability to
coppice strongly, they present a range of weeding situations requiring a variety of control methods.

Herbicides recommended are:

ROUNDUP (glyphosate)—foliar spray, stem treatment,
cut stump;
TIMBREL (triclopyr)—foliar spray, stem treatment,
cut stump;
AMCIDE (ammonium sulphamate)—stem treatment,
cut stump.

7.2 Types of treatment

Control methods fall into three groups:

Foliar treatment

This is normally the preferred method of application
wherever the weeds are in leaf and the foliage is accessible
to be sprayed. Weeds must however be small enough to
allow an effective dose of herbicide to be applied to the foliage. When treating small weeds less than 1 metre tall it is
practicable to consider application rates on a per treated
hectare basis. But when weeds are tall and are growing in
clumps or bushes it is more practicable to consider applying a herbicide solution of a given strength to wet all the
foliage of the weed species. Tall stems which need to be cut

and cannot for some reason be stump treated may be allowed to regrow for 1 or 2 years before spraying regrowth. Table 2 gives an indication of the susceptibilities of the various species to the recommended herbicides.

Table 2 Product rates for woody species susceptible to foliar application

Species	Herbicide rate (litres/ha)	
	Roundup*	Timbrel
Acer spp. (sycamore, maple)	3	4
Alnus spp. (alder)	—	4
Betula spp. (birch)	2	4
Castanea sativa (chestnut)	—	6
Cornus sanguinea (dogwood)	—	4
Corylus avellana (hazel)	3	6
Crataegus monogyna (hawthorn)	3	MR**
Cytisus scoparius (broom)	MR	2
Fagus sylvatica (beech)	2	6
Fraxinus excelsior (ash)	3	8
Ligustrum vulgare (privet)	2	6
Populus tremula (aspen)	2	4
Prunus laurocerasus (laurel)	8–10	8
Prunus spinosa (blackthorn)	2	4
Quercus spp. (oak)	3	8
Rhamnus cathartica (buckthorn)	—	6
Rhododendron ponticum (rhododendron)	8–10	8
Rosa canina etc. (wild rose)	2	2
Rubus spp. (bramble)	3	2
Salix spp. (willow)	3	6
Sambucus nigra (elder)	2	4
Sorbus aucuparia (rowan)	2	6
Ulex spp. (gorse)	R/MR	2
Ulmus procera (elm)	2	6
Viburnum opulus (guelder rose)	5	4

Notes:
*   —For preplant treatments, particularly in areas with dense growth or moderately resistant species the rate should be increased to 5 litres per hectare to ensure effective control.
**   —Susceptible to stump treatment see Section 7.5.2.
MR = moderately resistant.
R   = resistant.
—   = not tested.

Stem treatments

The techniques of frill girdling and stem injection are
appropriate for the treatment of woody growth that is too
tall to allow the foliage to be sprayed effectively and where
the resulting dead standing stems can be accepted or
conveniently removed. Costs can be high if there is a high
stocking of unwanted woody growth.

Cut stump treatment

This method is used for control of coppice regrowth after
felling (of crop trees or scrub) and avoids the problem of
unsightly dead stems remaining standing on the site. As
with stem treatments, the cost per treated hectare can be
high if there is a high stocking of unwanted woody growth.

7

7.3   *Foliar treatment*

7.3.1   GLYPHOSATE

Approved product
ROUNDUP 360 g/litre glyphosate (Monsanto).

Description
A translocated herbicide taken up by the foliage and conveyed to the roots. It causes chlorosis and eventual death of leaves and kills roots and shoots.

Roundup controls a wide range of weeds including grasses, herbaceous broadleaved weeds, bracken, heather and woody weeds. When applied late in the growing season, the main effect is obtained in the following year when roots die and suckering is prevented.

On contact with the soil, Roundup is inactivated and quickly broken down. Planting can be carried out immediately after treatment, but a minimum of 5 days should be allowed before cultivation and the breaking up of rhizomes and roots.

Crop tolerance
SS, NS, SP, CP, LP, RC and LC: will tolerate overall sprays provided leader growth has hardened. Hardening can occur as early as the end of July or may be delayed until the end of October in some locations and seasons. To avoid damage to lammas growth, herbicide sprays must be directed away from leaders. During the active growing season trees must be guarded or the spray placed to avoid contact with the crop.

DF and NF: as above but much more sensitive.

Broadleaved trees, larch and other conifers will not tolerate overall applications: always use a guard, a weedwiper or a directed spray to avoid contact with the foliage and immature bark of crop trees.

Product rate
Apply 2.0–5.0 litres (see Table 2) of product per treated hectare diluted in water and sprayed to wet all foliage.

Method of application
Pre-plant (overall application) or post-plant (overall or directed application)
Tractor mounted equipment
Boom sprayer at LV or MV.
Ulvaforest low speed rotary atomiser at VLV.

Handheld equipment
Knapsack sprayer at MV.
Knapsack sprayer with 'VLV' nozzle at LV.
Forestry Spot Gun for directed application.
Mistblower at LV. This method is more effective than ULV for taller denser foliage as the fan-assisted flow gives better penetration.
ULVA high speed rotary atomiser at VLV. For full effect, dilute the herbicide with at least 5 times the volume of water.

Post-plant (directed application)
Handheld equipment
Knapsack sprayer (with guard if required) at MV or with 'VLV' nozzle at LV.
Forestry Spot Gun for directed application.

Refer to Section 11 for details of applicator and correct calibration.

Timing of application
June to August inclusive but after new growth on crop trees has hardened (see Crop tolerance). This is also the optimum time for pre-planting sprays of woody regrowth.

Additional information
1. Weed control
a. Roundup applied later than June will be too late to lessen the effect of weed competition in the current season.
b. Roundup is most effective on moist vegetation and when relative humidity is high and the air is warm (e.g. 15°C+).
c. Reduced weed control may result when weeds are under stress, e.g. frost or drought.

d. On woody weeds and mixtures, mistblowing is usually the most effective of the handheld methods but it is also likely to inflict a higher degree of damage on unhardened crop trees.

e. Heavy rainfall within 24 hours of application may reduce the herbicide's effectiveness by preventing sufficient foliar absorption. The addition of Mixture B at 2% of final spray volume will improve reliability in these circumstances. The addition of Mixture B will reduce crop tolerance and it must only be used pre-plant or post-plant as a directed spray.

f. Diluted Roundup may denature after 2 to 3 days. Where possible use tap water as the diluent and only mix sufficient for the day's programme.

2. *Protective clothing*

Read product label for protective clothing and equipment requirements and check there are no items required in addition to the Forestry Commission recommendations in Section 10.

7

3. *Special precautions*

Roundup concentrate can be harmful to fish; do not contaminate ponds, waterways and ditches with chemical or used containers.

The label on the herbicide container has been designed for your protection—**ALWAYS READ THE INSTRUCTIONS ON THE LABEL.**

7.3.2   *TRICLOPYR*

Approved product
TIMBREL 480 g/litre triclopyr (Dow Agriculture).

Description
A plant growth regulating herbicide which is rapidly absorbed mainly through the foliage but also by roots and stems; once inside the plant it is readily translocated. Timbrel is effective against a wide range of weed species (see Table 2).

Grasses sometimes show some yellowing following spraying operations but this is quickly outgrown.

In the soil Timbrel is broken down fairly rapidly by microbial action. Planting should be deferred for at least 6 weeks after application.

Crop tolerance
Spruces: will tolerate overall sprays up to 4 l/ha provided leader growth has hardened. Hardening can occur as early as the end of July or may be delayed until October in some locations and seasons. To avoid damage to lammas growth, herbicide sprays must be directed away from leaders. During the active growing season trees must be guarded or the spray placed to avoid contact with the crop.

Pines: rather more sensitive than spruces with occasional leader damage if sprayed overall.

Larch, other conifers and broadleaves: severely damaged by overall sprays while in leaf. Late September application will be tolerated if applied with great care to avoid the foliage of crop trees.

Product rate
Apply 2.0—8.0 litres (see Table 2) of product per treated hectare diluted in water and sprayed to wetness.

Methods of application
Pre-plant (overall application)
Tractor mounted equipment
Tractor mounted boom sprayer at MV.
Handheld equipment
Knapsack sprayer at MV.

Post-plant (directed or overall application)
Handheld equipment
Knapsack sprayer (with guard if required) at MV.
Forestry Spot Gun.

Refer to Section 11 for details of applicator and correct calibration.

Timing of application
The timing of overall applications will be determined by the growth stage of the crop trees (see Crop tolerance), but to be effective the operation must take place after crop trees have hardened but before the target species show signs of senescence.

Directed applications can take place at any time between June—September provided the target species shows no signs of senescence and adequate precautions are taken to protect the crop species.

Additional information
1. Weed control
a. If Timbrel is applied during very hot weather some volatilisation may occur with consequent risk to sensitive crops downwind.
b. Rainfall within 2 hours of application may reduce the herbicide's effectiveness by preventing sufficient foliar absorption.
c. A minimum interval of 6 weeks is required between application of Timbrel and planting.

2. Protective clothing
Read product label for protective clothing and equipment requirements and check there are no items required in addition to the Forestry Commission recommendations in Section 10.

3. Special precautions
a. Timbrel is a mild eye irritant.
b. Timbrel should not be applied via controlled droplet applicators.

c. Timbrel is dangerous to fish: do not contaminate ponds, waterways and ditches with concentrate, spray drift or used containers.

The label on the herbicide container has been designed for your protection—**ALWAYS READ THE INSTRUCTIONS ON THE LABEL.**

7

7.4 Stem treatments

7.4.1 GLYPHOSATE

Approved product
ROUNDUP 360 g/litre glyphosate (Monsanto).

Description
A translocated herbicide normally applied to and taken up by the foliage but which is also effective as a stem injection treatment.

It causes chlorosis and eventual death of leaves and kills roots and shoots.

Stem injection with Roundup controls all the major broadleaved woody weed species and is also highly effective in killing individual stems of SS and other conifers, e.g. in chemical thinning.

Crop tolerance
There is no evidence of translocation across root grafts to untreated trees ('flashback'). Unwanted stems can be safely treated by this method among any crop species.

For foliar crop tolerance see Section 7.3.1.

For treatment of stems with a cleaning saw and application of Roundup via a cleaning saw attachment to the cut stump, see Section 7.5.1.

Product rate
Apply 2 ml of a 50% solution of Roundup in water per 10 cm stem diameter.

For trees with a diameter greater than 10 cm, separate incisions evenly spaced around the stem should be made for each 10 cm stem diameter.

Method of application
Application should be made using either:
a. downward axe strokes to penetrate the live cambial tissue and then applying Roundup into the notches using a Forestry Spot Gun;
b. suitable tree injection equipment (Jim Gem or Hypo Hatchet). Incisions should be at least 2.5 cm wide and sited between ground level and 1 metre high.

For trees up to 10 cm diameter: cut a single notch on one side of the stem. For larger trees: a second cut should be made on the opposite side of the stem.

Timing of application
At any time of year except during the period of maximum sap flow in spring. This usually occurs during March—May.

Additional information
1. *Weed control*
a. Roundup applied later than June will be too late to lessen the effect of weed competition in the current season.
b. Diluted Roundup may denature after 2 to 3 days. Where possible use tap water as the diluent and only mix sufficient for the day's programme.

2. *Protective clothing*
Read product label for protective clothing and equipment requirements and check there are no items required in addition to the Forestry Commission recommendations in Section 10.

3. *Special precautions*
Roundup concentrate can be harmful to fish; do not contaminate ponds, waterways and ditches with chemical or used containers.

The label on the herbicide container has been designed for your protection—ALWAYS READ THE INSTRUCTIONS ON THE LABEL.

7.4.2 *TRICLOPYR*

Approved product
TIMBREL 480 g/litre triclopyr (Dow Agriculture).

Description
A plant growth regulating herbicide which is rapidly
absorbed mainly through the foliage but also by roots and
stems; once inside the plant it is readily translocated.
Timbrel is effective against a wide range of weed species
but hawthorn is relatively resistant. Grasses sometimes
show some yellowing following spraying operations but this
is quickly outgrown.
   In the soil Timbrel is broken down fairly rapidly by
microbial action. Planting should be deferred for at least 6
weeks after application.

Crop tolerance
There is no evidence of translocation across root grafts to
untreated trees ('flashback'). Unwanted stems can be safely
treated by this method among any crop species.
   For foliar crop tolerance see Section 7.3.2.

Product rate
Apply 2 ml undiluted or 4 ml diluted 1:1 with water per
5 cm stem diameter.
   For trees with trunks larger than 5 cm diameter,
separate incisions evenly spaced around the stem should be
made for each 5 cm stem diameter.

Method of application
Application should be made using either:
a. downward axe strokes to penetrate the live cambial
   tissue and then applying Timbrel into the notches using
   a Forestry Spot Gun;
b. suitable tree injection equipment (Jim Gem or Hypo
   Hatchet).
   Incisions should be at least 2.5 cm wide and sited between ground level and 1 metre high.

Timing of application
Applications may be made at any time of year but best
results follow summer or autumn treatments.

Additional information

1. Weed control
If Timbrel is applied during very hot weather some volatilisation may occur with consequent risk to sensitive crops
downwind.

2. Protective clothing
Read product label for protective clothing and equipment
requirements and check there are no items required in
addition to the Forestry Commission recommendations in
Section 10.

3. Special precautions
a. Timbrel is a mild eye irritant.
b. Timbrel is dangerous to fish; do not contaminate ponds,
   waterways and ditches with concentrate, spray drift or
   used containers.

  The label on the herbicide container has been designed
for your protection—**ALWAYS READ THE
INSTRUCTIONS ON THE LABEL.**

7

7.4.3 *AMMONIUM SULPHAMATE*

Approved product
AMCIDE soluble crystals; 100% ammonium sulphamate
(Battle, Hayward and Bower).

Description
A highly soluble translocated, contact and soil-acting herbicide which is absorbed through leaves roots and exposed
live tissue surfaces. It is effective against most woody
species including the more resistant species such as
rhododendron, hawthorn and ash. It corrodes metals and
alloys including copper, brass, mild steel and galvanised
iron.

Breakdown in the soil can take up to 12 weeks during
which time it retains its herbicidal properties. Three
months should elapse between treatment and subsequent
planting.

Crop tolerance
All crop species are severely damaged or killed by direct
application of Amcide or by absorption via the roots.

Post-planting frill-girding or notching can be carried out
safely if great care is taken to avoid unnecessary spillage
or overflow reaching the ground. Pre-planting treatment is
preferable.

Product rates and methods of application
Frill girdling
A frill is cut round each stem by overlapping downward
strokes of a light axe or billhook and exposed live tissue is
sprayed to runoff with a 40% solution of Amcide (0.4 kg
crystals per 1 litre of water) using a plastic watering can or
a Tecnoma T16 semi-pressurised knapsack sprayer (see
Section 11.5.2).

Notching
Individual notches are cut round the stem by using single
downward axe strokes penetrating the live cambial tissue.
Notches should be at least 3 cm long and should not be
further than 10 cm apart, edge to edge. 15 g of dry crystals
are placed in each notch. They should be sited between
ground level and 1 metre high.

7

Timing of application

Best results are obtained from applications made during the growing season but Amcide can be applied at any time between April and September.

Additional information

1. Weed control

a. Amcide is best applied in dry weather so that the spray solution or the crystals are not washed out of the stem frill or notch.

b. Solutions of Amcide in water should be freshly prepared on each day of use.

2. Protective clothing

Read product label for protective clothing and equipment requirements and check there are no items required in addition to the Forestry Commission recommendations in Section 10.

3. Special precautions

Amcide is harmful to fish; do not contaminate ponds, waterways or ditches with chemical or used container.

The label on the herbicide container has been designed for your protection—**ALWAYS READ THE INSTRUCTIONS ON THE LABEL.**

7

7.5 *Cut stump treatment*

7.5.1 *GLYPHOSATE*

Approved product
ROUNDUP 360 g/litre glyphosate (Monsanto).

Description
A translocated herbicide normally applied to and taken up
by the foliage but which is also effective when applied to
freshly cut stumps. As a cut stump treatment it reduces or
stops the production of new shoots and kills the root
system of the stump.

Roundup controls all the major broadleaved woody weed
species and is also highly effective in preventing regrowth
from remaining live branches on the stumps of SS and
other conifers, e.g. for respacement of natural regeneration.

Planting can be carried out immediately after treatment.

Crop tolerance
There is no evidence of translocation across root grafts to
untreated trees ('flashback'). Cut stumps can be safely
treated by this method among any crop species provided
none of the herbicide is allowed to fall on the crop foliage.

For foliar crop tolerance see Section 7.3.1.

Product rate
Conifer species and rhododendron: apply a 20% solution of
the product in water.

Broadleaved species: apply a 10% solution of the product
in water.

Method of application
Apply to saturate the freshly cut stump by:
knapsack sprayer operated at low pressure;
Forestry Spot Gun fitted with a solid stream nozzle;
paintbrush;
cleaning saw fitted with a suitable herbicide spray
attachment.

To aid identification of treated stumps a suitable dye
(e.g. from Hortichem Ltd.) may be added.

Timing of application

Best results are obtained when application is made to freshly cut stumps from October to March (outside the time of spring sap flow).

Application should be made within 1 week of felling.

Additional information

1. Weed control

a. Reduced weed control may result when Roundup is applied to the surface of frozen stumps.

b. Heavy rainfall within 24 hours of application may reduce the herbicide's effectiveness by preventing sufficient absorption.

c. Diluted Roundup may denature after 2 to 3 days. Where possible use tap water as the diluent and only mix sufficient for the day's programme.

2. Protective clothing

Read product label for protective clothing and equipment requirements and check there are no items required in addition to the Forestry Commission recommendations in Section 10.

3. Special precautions

Roundup concentrate can be harmful to fish: do not contaminate ponds, waterways and ditches with chemical or used containers.

The label on the herbicide container has been designed for your protection—ALWAYS READ THE INSTRUCTIONS ON THE LABEL.

7.5.2   *TRICLOPYR*

Approved product
TIMBREL 480 g/litre triclopyr (Dow Agriculture).

Description
A plant growth regulating herbicide which is rapidly absorbed mainly through the foliage but also by roots and stems; once inside the plant it is readily translocated.
Timbrel is effective against a wide range of weed species.

Grasses sometimes show some yellowing following spraying operations but this is quickly outgrown.

In the soil Timbrel is broken down fairly rapidly by microbial action.

Crop tolerance
There is no evidence of translocation across root grafts to untreated trees ('flashback'). Unwanted stems can be safely treated by this method among any crop species provided none of the herbicide is allowed to fall on the crop foliage.

For foliar crop tolerance see Section 7.3.2.

Product rate
Solution of product in water applied to the freshly cut stump
For alder, birch, blackthorn, box and dogwood use a 2% solution (20 ml/l).

For hawthorn, laurel and rhododendron use an 8% solution.

For all other broadleaved tree species, use a 4% solution.

Method of application
Apply to saturate the freshly cut surface of the stump and any remaining bark by:
knapsack sprayer operated at low pressure;
Forestry Spot Gun fitted with a solid stream jet;
paintbrush.

So that treated stumps can be identified during treatment a suitable dye (e.g. from Hortichem Ltd.) can be added to the spray solution.

Timing of application

Best results are obtained when application is made to freshly cut stumps from October to March (outside the time of spring sap flow).

Application should be made within 1 week of felling.

Additional information

1. Weed control

If Timbrel is applied during very hot weather some volatilisation may occur with consequent risk to sensitive crops downwind.

2. Protective clothing

Read product label for protective clothing and equipment requirements and check there are no items required in addition to the Forestry Commission recommendations in Section 10.

3. Special precautions

a. Timbrel is a mild eye irritant.
b. Timbrel is dangerous to fish; do not contaminate ponds waterways and ditches with concentrate, spray drift or used containers.

The label on the herbicide container has been designed for your protection—**ALWAYS READ THE INSTRUCTIONS ON THE LABEL**.

7.5.3 *AMMONIUM SULPHAMATE*

Approved product
AMCIDE soluble crystals; 100% ammonium sulphamate
(Battle, Hayward and Bower).

Description
A highly soluble translocated, contact and soil-acting herbicide which is absorbed through leaves roots and exposed
live tissue surfaces. It is effective against most woody
species including the more resistant species such as
rhododendron, hawthorn and ash. It corrodes metals and
alloys including copper, brass, mild steel and galvanised
iron.

Breakdown in the soil can take up to 12 weeks during
which time it retains its herbicidal properties. Three
months should elapse between treatment and subsequent
planting.

Crop tolerance
All crop species are severely damaged or killed by direct
application of Amcide or by direct or indirect root-
poisoning by percolation of the herbicide into the soil from
treated stems and stumps. Cut stump application of
Amcide should therefore be restricted to pre-planting.

Product rate and methods of application
Apply to freshly cut stump surfaces
a. a 40% solution of Amcide (0.4 kg crystals per 1 litre of
   water) to the point of runoff, using a plastic watering
   can or a Tecnoma T16 semi-pressurised knapsack
   sprayer (see Section 11.5.2) or
b. dry Amcide crystals at the rate of 6 g per 10 cm of
   stump diameter.

Timing of application
Best results are obtained from applications made during
June—September but Amide can be applied at any time
between April and September.

Additional information

1. Weed control

a. Amcide is best applied in dry weather so that the spray solution or the crystals are not washed off the treated stump.

b. Solutions of Amcide in water should be freshly prepared on each day of use.

2. Protective clothing

Read product label for protective clothing and equipment requirements and check there are no items required in addition to the Forestry Commission recommendations in Section 10.

3. Special precautions

Amcide is harmful to fish; do not contaminate ponds, waterways or ditches with chemical or used container.

The label on the herbicide container has been designed for your protection—**ALWAYS READ THE INSTRUCTIONS ON THE LABEL.**

7

8 Gorse and broom

8.1 *General*

These two evergreen shrubs occur separately or together and locally may present a major weed problem.

Recommended herbicide: TIMBREL (triclopyr).

8

8.2 *TRICLOPYR*

Approved product
TIMBREL 480 g/litre triclopyr (Dow Agriculture).

Description
A plant growth regulating herbicide which is rapidly
absorbed mainly through the foliage but also by roots and
stems; once inside the plant it is readily translocated.

Grasses sometimes show some yellowing following
spraying operations but this is quickly outgrown.

Timbrel is particularly effective as a foliar spray on
gorse and broom.

In the soil Timbrel is broken down fairly rapidly by
microbial action.

Planting should be deferred for at least 6 weeks after
application.

Crop tolerance
Spruces: will tolerate overall sprays provided leader growth
has hardened. Hardening can occur as early as the end of
July or may be delayed until the end of October in some
locations and seasons. To avoid damage to lammas growth,
herbicide sprays must be directed away from leaders.
During the active growing season trees must be guarded or
the spray placed to avoid contact with the crop.

Pines: rather more sensitive than spruces with occasional
leader damage, if sprayed overall.

Larch, other conifers and broadleaves: severely damaged
by overall sprays while in leaf. Late September application
will be tolerated if applied with great care to avoid the
foliage of crop trees.

Product rate
Apply as 2.0 litres of product per treated hectare in water.

Methods of application
*Pre-plant (overall application) or post-plant (directed,
overall or band application)*

Tractor mounted equipment
Tractor mounted boom sprayer at MV.

8

Handheld equipment
Knapsack sprayer at MV.

Spot application will rarely be appropriate among such aggressive weeds.

Refer to Section 11 for details of applicator and correct calibration.

Timing of application
Applications made April—October give best results; for post plant sprays see Crop tolerance above.

Additional information
1. Weed control
a. If Timbrel is applied during very hot weather some volatilisation may occur with consequent risk to sensitive crops downwind.
b. Rainfall within 2 hours of application may reduce the herbicide's effectiveness by preventing sufficient foliar absorption.
c. A minimum interval of 6 weeks is required between application of Timbrel and planting.

2. Protective clothing
Read product label for protective clothing and equipment requirements and check there are no items required in addition to the Forestry Commission recommendations in Section 10.

3. Special precautions
a. Timbrel is a mild eye irritant.
b. Timbrel should not be applied via controlled droplet applicators.
c. Timbrel is dangerous to fish: do not contaminate ponds, waterways and ditches with concentrate, spray drift or used containers.

The label on the herbicide container has been designed for your protection—ALWAYS READ THE INSTRUCTIONS ON THE LABEL.

8

9 Rhododendron

9.1 *General*

The glossy evergreen leaves of rhododendron and laurel
have a thick waxy cuticle which is comparatively resistant
to the entry of herbicides. To achieve adequate control
application rates need to be higher than for most other
woody weeds. Moreover in older bushes translocation is
very restricted and it is necessary to spray almost the
whole of the leaf area and stem surfaces to achieve good
control.

Rhododendron is to be found on acid sites, mainly in the
wetter western half of the country, in all phases of
colonisation from a light scatter of small seedlings, through
a partial cover of bushes to impenetrable thickets 2—5 m in
height.

The early stages of encroachment are (subject to terrain)
easily accessible for herbicide application but bushes more
than 1.5 m high must be manually or mechanically cleared
to allow the stumps and the more susceptible regrowth to
be sprayed, preferably before the regrowth is more than
1 m tall.

Recommended herbicides are: ROUNDUP (glyphosate);
TIMBREL (triclopyr);
AMCIDE (ammonium
sulphamate).

These herbicides are applied either as a foliar spray or as
a cut stump treatment, and frequently a combination of
both techniques is used where cut stumps and young
regrowth exist together.

9.2 *Glyphosate*

Approved product
ROUNDUP 360 g/litre glyphosate (Monsanto).

Description
A translocated herbicide taken up by the foliage and conveyed to the roots. It causes chlorosis and eventual death of leaves and kills roots and shoots.

With most species, once inside the plant it is readily translocated, but translocation within rhododendron seems to be particularly poor in a tangential direction, so that spraying part of a bush results in the death of that part only.

Roundup controls a wide range of weeds including grasses, herbaceous broadleaved weeds, bracken, heather and woody weeds. With the latter group there may be little effect until the following season when roots are killed and much resuckering prevented.

On contact with the soil Roundup is inactivated and quickly broken down. Planting can be carried out immediately after treatment, but a minimum of 5 days should be allowed before cultivation and the breaking up of rhizomes and roots.

9

Crop tolerance
At the rates of application needed to kill rhododendron, all forestry crop species are severely damaged by overall treatments of Roundup. Ideally all rhododendron control should be carried out pre-planting but, if a crop is present, treatment is possible if the spray is very carefully directed to avoid contact with crop trees.

Product rate
Apply 10 litres of Roundup per hectare
> *or* 8 litres of Roundup per hectare plus Mixture B at 2% of final spray volume.

Alternatively spray so that all the foliage is wetted, but the herbicide solution does not run off, with a 2% (2 litres of product in 100 litres water) solution using a knapsack sprayer or a mistblower as appropriate.

For cut stump treatment follow coniferous recommendation, see Section 7.5.1.

Method of application
Pre-plant (overall application)
Tractor mounted equipment
Boom sprayer at LV or MV.
Ulvaforest low speed rotary atomiser at VLV.

Handheld equipment
Knapsack sprayer at MV.
Knapsack sprayer with 'VLV' nozzle at LV.
Herbi low speed rotary atomiser or ULVA high speed
rotary atomiser at VLV: for full effect dilute the herbicide
with at least 3 times the volume of water.
Mistblower at LV. This method is very effective as the fan-
assisted flow gives good penetration, thus fewer access
racks may be necessary.

Post-plant (directed)
Handheld equipment
Knapsack sprayer at MV.
Knapsack sprayer with 'VLV' nozzle at LV.
Herbi low speed rotary atomiser at VLV.
The Forestry Spot Gun will only be appropriate for
treating seedling rhododendron.

Refer to Section 11 for details of applicator and correct
calibration.

Timing of application
June to September.

Additional information
1. Weed control
a. Roundup applied later than June will be too late to
   lessen the effect of weed competition in the current
   season.
b. Roundup is most effective on moist vegetation and when
   relative humidity is high and the air is warm (e.g.
   15°C+).
c. Reduced weed control may result when plants are under
   stress, e.g. drought.
d. Heavy rainfall within 24 hours of application may reduce
   the herbicide's effectiveness by preventing sufficient
   foliar absorption. The addition of Mixture B at 2% of
   final spray volume will improve reliability in these

circumstances. The addition of Mixture B will reduce crop tolerance and it must only be used pre-plant or post-plant as a directed spray.
e. Diluted Roundup may denature after 2 to 3 days. Where possible use tap water as the diluent and only mix sufficient for the day's programme.

2. Protective clothing
Read product label for protective clothing and equipment requirements and check there are no items required in addition to the Forestry Commission recommendations in Section 10.

3. Special precautions
Roundup concentrate can be harmful to fish: do not contaminate ponds, waterways and ditches with chemical or used containers.

The label on the herbicide container has been designed for your protection—**ALWAYS READ THE INSTRUCTIONS ON THE LABEL.**

9

9.3 TRICLOPYR

Approved product
TIMBREL 480 g/litre triclopyr (Dow Agriculture).

Description
A plant growth regulating herbicide which is rapidly absorbed mainly through the foliage but also by roots and stems. With most species, once inside the plant it is readily translocated, but translocation within rhododendron seems to be particularly poor in a tangential direction, so that spraying part of a bush results in the death of that part only.

Grasses sometimes show some yellowing following spraying operations but this is quickly outgrown.

In the soil Timbrel is broken down fairly rapidly by microbial action.

Planting should be deferred for at least 6 weeks after application.

Crop tolerance
At the rates of application needed to kill rhododendron all crop species are likely to be damaged by overall treatments of Timbrel. Ideally all rhododendron control should be carried out pre-planting but, if a crop is present, treatment is possible if the spray is very carefully directed to avoid contact with crop trees.

Product rate
Apply 8 litres of product per treated hectare in water.

Alternatively, for taller denser rhododendron, spray so that all the foliage is wetted but the herbicide does not run off, using a mixture of 130 ml Timbrel in 5 litres of water (130 ml Timbrel + 4870 ml water).

Method of application
Tractor mounted equipment
Tractor mounted boom sprayer at MV.
Handheld equipment
Knapsack sprayer at MV.

The Forestry Spot Gun will only be appropriate for treating seedling rhododendron.

Refer to Section 11 for details of applicator and correct calibration.

Timing of application
June—September.

Additional information
1. Weed control
a. If Timbrel is applied during very hot weather some volatilisation may occur with consequent risk to sensitive crops downwind.
b. Rainfall within 2 hours of application may reduce the herbicide's effectiveness by preventing sufficient foliar absorption.
c. A minimum interval of 6 weeks is required between application of Timbrel and planting.

2. Protective clothing
Read product label for protective clothing and equipment requirements and check there are no items required in addition to the Forestry Commission recommendations in Section 10.

3. Special precautions
a. Timbrel is a mild eye irritant.
b. Timbrel should not be applied via controlled droplet applicators.
c. Timbrel is dangerous to fish: do not contaminate ponds, waterways and ditches with concentrate, spray drift or used containers.

The label on the herbicide container has been designed for your protection—ALWAYS READ THE INSTRUCTIONS ON THE LABEL.

9

9.4 *AMMONIUM SULPHAMATE*

Approved product
AMCIDE soluble crystals; 100% ammonium sulphamate
(Battle, Hayward and Bower).

Description
A highly soluble translocated, contact and soil-acting
herbicide which is absorbed through leaves, roots and
exposed live tissue surfaces. It is effective against most
woody species including the more resistant species such as
rhododendron, hawthorn and ash. It corrodes metals and
alloys including copper, brass, mild steel and galvanised
iron.

Breakdown in the soil can take up to 12 weeks during
which time it retains its herbicidal properties. Three
months should elapse between treatment and subsequent
planting.

Crop tolerance
All crop species are severely damaged or killed by direct
application of Amcide or by absorption via the roots.
Amcide should therefore only be used pre-planting.

Product rates and methods of application
Apply a 40% solution of Amcide (0.4 kg crystals per 1 litre
of water) using a plastic watering can or a Tecnoma T16
semi-pressurised knapsack sprayer (see Section 11.5.2) to all
accessible surfaces including freshly cut stumps and all
remaining bark, twigs and foliage, wetting to the point of
run-off.

Timing of application
Best results are obtained from applications made during
the growing season with optimum control achieved with
May and June treatments. But Amcide can be applied at
any time between April and September.

Additional information
1. Weed control
a. Amcide is best applied in dry weather so that the spray
   solution is not washed off treated surfaces.

b. Solutions of Amcide in water should be freshly prepared on each day of use.

2. *Protective clothing*
Read product label for protective clothing and equipment requirements and check there are no items required in addition to the Forestry Commission recommendations in Section 10.

3. *Special precautions*
Amcide is harmful to fish: do not contaminate ponds, waterways or ditches with chemical or used containers.

The label on the herbicide container has been designed for your protection—**ALWAYS READ THE INSTRUCTIONS ON THE LABEL.**

9

10 Protective clothing and personal equipment

10.1 *General*

The appropriate protective equipment, listed in Table 3, should be made available on a personal basis to all users of herbicides, including those handling herbicide containers. All protective equipment should be kept clean and in good repair. Any damaged item should be replaced promptly.

To avoid contamination of personal clothing or skin it is essential that the contaminated outside of the clothing does not come into contact with the clean inside during transit or when putting on or removing protective clothing and equipment.

The order for removing clothing and equipment is:

First wash gloves and then remove in the following order,

face shield
filtering facepiece respirator
wellington boots
impervious suit (either one or two piece)
gloves (after washing again to remove any
contamination picked up during clothing removal).

Clean clothing should be put on in reverse order without the necessity of washing gloves.

At the end of the day all clothing and equipment should be thoroughly washed down, dried and left to hang overnight.

At the end of the spraying season, or more frequently if necessary, items should be thoroughly washed and dried either at the forest store or by a reputable cleaning company. Items must never be washed with normal domestic washing. Cleaned items should be hung up and stored in a cool dry place away from direct sunlight, vermin and pesticides.

Contaminated items of protective clothing and equipment must not be worn or carried inside the cab of any vehicle.

Always read the product label in case there is need for additonal items of protective clothing or equipment to that recommended in Table 3.

10.2   *List of recommended products and suppliers*

Equipment	Recommendations	Suppliers (see footnote)
Wellington boots	Dunlop Safety 8807 or 8808 with steel midsole	Greenham Tool Co. Ltd. 671 London Road Isleworth Middlesex TW7 4EX
	Suretred Safety 881 or 882 with steel midsole	Frank Parker and Co. Ltd. Whitehouse Road Stirling FK7 7SJ
Shoe chains	Rudd Shoe Chains Size 1: Shoe sizes under 5 Size 2: Shoe sizes 5–9 Size 3: Shoe sizes over 9	Rudd Chains Ltd. 1–3 Belmont Road Whitstable Kent CT5 1QT
Impervious suit (trousers & jacket with hood)	Gore-Tex Suits in small, medium & large sizes	Edward MacBean and Co. Ltd. 1–7 Napier Place Wardpark North Cumbernauld Glasgow G68 OLL (FC supplied via Blairadam clothing store. Ref PC 48).
Gloves	Edmont Solvex NX37–175 Length 30 cm sizes 7–11 It is recommended that two pairs of gloves are issued to each operator	Safety Specialist Ltd. Unit 8 Field House Industrial Estate Peter Street Sheffield S4 7SF
Cotton liner gloves (for use with the Forestry Spot Gun)	Mens Cotton Stockinette Gloves, knitted wrist: Code 304111	Greenham Tool Co. Ltd. 671 London Road Isleworth Middlesex TW7 4EX
Face shield	James North No. FS 1318 BW Shield depth 20 cm	Greenham Tool Co. Ltd. 671 London Road, Isleworth Middlesex TW7 4EX

10

Equipment	Recommendations	Suppliers (see footnote)
Face shield for use with safety helmet for drivers of tractors without cabs	FC H417 Aluminium frame for Safety Helmet FL 8 PC (20cm) clear polycarbonate screen wide flare	Protective Safety Ltd. Great George Street Wigan Greater Manchester WN3 4DE
Filtering face-piece respirator	3 M's 8810 Generally requires replacing after 2 hours use on a misty day or every 4 hours on a dry day	Herts Packaging Co. Ltd. 29 Mill Lane Welwyn Herts AL6 9EU

Personal hygiene

Barrier cream	Rozalex Wet Guard available in 450 ml containers	Lever International Ltd. PO Box 208 Lever House St. James Road Kingston-upon-Thames Surrey KT1 2BB
Skin cleanser for use with water	Arrow Chemicals Tuffstuff available in 1 litre dispenser bottle or 5, 10 and 20 litre containers	The Arrow Chemical Group of Companies PO Box 3 Stanhope Road Swadlincote Near Burton-on-Trent Staffs DE11 9BE
Paper towels	Kimwipes Steel Blue 10" rolls, Code 7148 In a case of 24 rolls	Kimberley Clark Ltd. Industrial Division Larkfield Kent ME20 7PS (Private sector: purchase locally from any industrial clothing supplier)

Note: This list is included as a guide to sources of supply and is **NOT** a comprehensive compilation of suppliers of recommended products. The omission of names of other possible suppliers does not imply that their services are unsatisfactory.

10.3 *Cleaning recommendations*

Any item of protective clothing or equipment which becomes grossly contaminated should be immediately rinsed thoroughly with water.

Wellington boots

At the end of each day wash down outsides of boots in a mild dilute solution of detergent and rinse in clean water. Allow both inside and outside to dry.

Gloves

Gloves should be washed:

- at the end of each work period, before removing protective clothing;
- after removing protective clothing;
- at the end of each working day.

Face shield

Wipe down with a mild dilute solution of detergent and allow to dry.

Impervious coverall (Gore-Tex jacket and trousers)

At the end of each day wash down with a mild dilute solution of a household washing detergent (not washing up liquid), and rinse in clean water. Allow both the inside and outside to dry.

If detergents are used to wash the neoprene part of the trouser legs after spraying Silvapron D a reaction can occur causing the neoprene to become swollen and tacky. To overcome this problem the manufacturers recommend that neoprene clothing is washed in clean water. Do not dry clean the suit.

Do not used high pressure hoses to wash down the suits.

When Gore-Tex suits are in use they should be handled with care as rough treatment will cause blistering of the Gore-Tex and reduce the liquid repellent properties.

To test the repellent properties of the Gore-Tex material, paper towelling should be placed inside the suit and water applied, without pressure, together with a light rubbing action to the outside of the garment for 5–10 seconds. If the paper becomes wet the suit will have lost its repellent properties and should be replaced.

10

10.4

Table 3 Protective clothing and equipment recommendations

Operation	Application equipment	Wellington boots	Impervious coverall			Gloves	Face shield	Filtering facepiece respirator	Ear defenders	Notes
			Trousers	Jacket	Hood					
Handling concentrate, mixing and filling (granules and liquid)	All types	E	E	E		E	E	—	—	
Application of granules or crystals	Gravity	E	E	D	D	E	D	—	—	
Medium volume (MV) and low volume (LV) spraying	Knapsack, nozzle height less than 1 metre	E	E	E	D	E	D	—	—	(1)(2)
	Knapsack, nozzle height greater than 1 metre	E	E	E	E	E	E	E	—	
	Forestry Spot Gun nozzle height less than 1 metre	E	E	D	D	E	D	—	—	(1)(2) (3)
	Forestry Spot Gun nozzle height greater than 1 metre	E	E	E	E	E	E	E	—	(3)
	Motorised mistblower	E	E	E	E	E	E	E	E	
Controlled droplet applicator (band/overall)	Microfit Herbi (low speed rotary atomiser)	E	E	E	D	E	D	—	—	

Controlled droplet applicator (incremental/placed)	Ulva 8 (high speed rotary atomiser)	E	E	E	E	E	E	—
Direct applicator	Weedwiper	E	E	D	D	D	—	—
Stem treatment tree injection	Hypo Hatchet/Forestry Spot Gun	E	E	E	E	E	—	—
	Jim Gem	E	E	D	D	D	—	—
Brushcutter stump treatment attachment	ENSO	E	E	D	D	D	—	—
Tractor mounted sprayers (MV, LV, VLV)	Tractor with fully enclosed cab	—	—	—	—	—	—	(4)
	Tractor without cab	E	E	E	E	D	E	(4)

Notes:

1. Face shield must be worn when applying Terbuthylazine with Atrazine.
2. Face shield and hood must be worn when applying Hexazinone
3. For operator comfort, cotton liner gloves may be worn during application.
4. When repairing or maintaining the sprayer the clothing requirements are as for handling, mixing and filling.

E = essential: either advised under Control of Pesticides Regulations 1986 and Control of Substances Hazardous to Health Regulations 1988 or considered necessary in relation to working condition.

D = discretionary: these items are not usually required but should be supplied on request to the operator or when noticeable exposure to a herbicide may arise through operations in unusual circumstances.

Recommendation for protective clothing and equipment requirements for use with application equipment not listed should be obtained from Eastern Region Work Study Team, Forestry Commission, Santon Downham, Brandon, Suffolk, IP27 OTJ. Telephone 0842 814227.

10

11 Equipment

List of contents

11

11.1 *Disclaimer*

The applicators detailed in this section are those which are currently recommended for use within the Forestry Commission; this does not imply that other manufacturers' applicators with similar capabilities and features are not suitable for applying the herbicides detailed in this Field Book.

11.2 *Volume rate categories*

High volume (HV) — over 700 litres per hectare
Not recommended for herbicide application except for cut stump spot treatment where high volume rate is recommended for optimum results. In all other situations high volume application can result in waste and ground contamination due to run off. Better results with practically no run off can be obtained by using lower volumes per hectare.

Applicators
Knapsack (cut stump application only).
Forestry Spot Gun (cut stump application only).

Medium volume (MV) — 200 to 700 litres per hectare
These rates give a good overall cover in most situations.

Applicators
Tractor mounted boom sprayer.
Knapsacks.

11

Low volume (LV) — 50 to 200 litres per hectare
Good coverage is achieved in pressure controlled sprayers by the use of precisely engineered hydraulic nozzles and by the production of small droplets with the mistblower.

Applicators
Tractor mounted boom sprayers fitted with LV nozzles.
Knapsacks with VLV nozzles.
Mistblowers.
Forestry Spot Gun (for use on grasses and herbaceous broadleaved weeds and for woody weed foliar application).

Very low volume (VLV) — 10 to 50 litres per hectare
Controlled droplet applicators (rotary atomisers) or pre-cisely engineered hydraulic nozzles must be used at these volumes to obtain a good droplet size spectrum and good droplet dispersal.

Applicators
Low speed rotary atomisers.
High speed rotary atomisers.
Forestry Spot Gun (for use on grasses and herbaceous broadleaved weeds).

Ultra low volume (ULV) — under 10 litres per hectare
Controlled droplet application (rotary atomiser) must be used to obtain adequate control of droplet size and droplet dispersal.

Applicator
High speed rotary atomisers.

11.3   *Calibration*

11.3.1   **Principles**
It is a requirement of the Control of Pesticides Regulations 1986 that maximum application rates are not exceeded. Overdosing with herbicide is expensive and places the oper-ator, the environment and crop trees at unnecessary risk. Underdosing can also be expensive if failure to control the weeds results in the need for subsequent operations. It is therefore desirable to achieve the target rate of application.

   To achieve the target rate of application distributed evenly over the target area it is necessary to determine the four parameters indicated below.

Volume rate (i.e. the quantity of spray mixture (diluent plus herbicide) applied in litres per hectare)
From the Method of application paragraph of the selected herbicide, determine the appropriate applicator and therefore the required volume rate.

11

Any volume rate within the range (i.e. MV 200–700) can be selected by varying the amount of diluent but keeping the rate at which the herbicide is applied constant.

Throughout this Field Book the rate of herbicide to be applied is given in litres or kilograms per treated hectare.

Speed (walking speed or tractor speed)

Walking speed

Measure the walking speed (in metres per minute), which can be sustained on site with the applicator full, and wearing the necessary protective clothing. Deviation from this speed will lead to over or under-dosing.

Tractor speed

This is determined by the roughness of the terrain.

Swathe width

Determine the swathe width (or spot diameter) required. Note that the height at which the nozzle is held has to be increased or decreased to take into account changes in vegetation height in order to maintain the determined swathe width at the main or average height of the target and avoid over or under dosing.

Nozzle output

Measure the nozzle output over a set period and convert to millilitres per minute. Note that changes in air temperature in the course of the day can alter the viscosity of the herbicide, requiring nozzle output to be remeasured.

11

11.3.2   **Calculation**

Band or overall application through the knapsack, Microfit Herbi, Herbi 77, ULVA or ULVAFOREST

Of the four parameters, nozzle output is generally the most appropriate to vary, either by adjusting the pressure or changing nozzles. 'Required nozzle output' can be calculated using the following equation:

$$\text{Required nozzle output (ml/min)} = \frac{\text{Walking speed (metres/min)} \times \text{Volume rate (litres/ha)} \times \text{Swathe width (metres)}}{10}$$

Example:
Required nozzle output for CP15 knapsack sprayer for application where:

> walking speed 45 m/min
> volume rate 150 l/ha (Section 4.5 Glyphosate)
> swathe width 1.2 m

$$\text{Required nozzle output} = \frac{45 \times 150 \times 1.2}{10} = 810 \text{ ml/min}$$

Select from the nozzle data of the CP15 specification a floodjet nozzle (see Section 11.5.1) which gives the nearest output to that calculated, in this case the VLV AN 2.0 Blue. Fit and adjust pressure until the required nozzle output is achieved.

Reducing the pressure reduces the nozzle output and produces larger droplets: increasing pressure increases the nozzle output and produces smaller droplets. A significant increase of pressure can produce small droplets which can drift off target, such increases should be avoided.

If the required nozzle output is not achievable with any nozzle the volume rate should be varied by increasing or decreasing the quantity of diluent. Gravity feed applicators (e.g. Herbi's, ULVAs) may, even after altering the volume rate, not achieve the required nozzle output; walking speed should then be considered.

Spot application through the knapsack or granule applicator
To calculate volume per spot the following equation has to be solved:

$$\begin{matrix}\text{Volume per spot} \\ \text{(ml or g)}\end{matrix} = \frac{\begin{matrix}\text{Volume rate} \\ \text{(litres or kg/ha)}\end{matrix} \times \begin{matrix}\text{Area of spot} \\ \text{(m}^2\text{)}\end{matrix}}{10}$$

Walking speed does not affect the achievement of target application rate, during spot treatment and can therefore be discarded for calculation purposes.

Example 1:
The Pepperpot or Moderne applying Atlas Lignum to a 1 metre diameter spot.
Volume Rate 40 kg/ha (Section 4.3 Atrazine with Dalapon)
Area of spot with a diameter of 1 m = 0.785 m^2

Therefore volume per spot $= \dfrac{40\times0.785}{10} = 3.14$ g

The operator should practise application over a paper or plastic sheet until the required area and volume per spot is achieved.

Example 2:

For spot application by the knapsack the timing of the dose required has to be determined, after the volume per spot has been calculated.

Volume rate 200 l/ha (Section 4.5 Glyphosate)

Area of spot with a diameter of 1.1 m=0.95 m^2. Therefore:

Volume per spot $= \dfrac{200\times0.95}{10} = 19$ ml

Select nozzle from the CP15 specification (Section 11.5.1). Divide the required volume per spot by the nozzle output to give a time in seconds required to treat a spot. The operator should practise application over a graduated measure until proficient at achieving the volume per spot and over a dry surface to achieve the required spot diameter.

The sections which deal with the individual applicators give detailed instructions on calibration.

For other calculations Section 11.3.3 lists useful equations and tables of the area of plantation treated when herbicides are applied as spot or band treatments.

Calibration for the Forestry Spot Gun and Weedwiper is detailed under their respective applicator sections.

11.3.3 **Calibration equations for calculating various parameters applicable to different types of applicators**

11

1. Nozzle output (ml/min) $= \dfrac{\substack{\text{Walking speed} \\ \text{(metres/min)}} \times \substack{\text{Volume rate} \\ \text{(litres/ha)}} \times \substack{\text{Swathe width} \\ \text{(metres)}}}{10}$

2. Walking speed (metres/min) $= \dfrac{\substack{\text{Nozzle output} \\ \text{(ml/min)}} \times 10}{\substack{\text{Volume rate} \\ \text{(litres/ha)}} \times \substack{\text{Swathe width} \\ \text{(metres)}}}$

3. $\text{Volume rate (litres/ha)} = \dfrac{\text{Nozzle output (ml/min)} \times 10}{\text{Walking speed (metres/min)} \times \text{Swathe width (metres)}}$

4. $\text{Distance covered when band spraying (metres)} = \dfrac{\text{Total volume of sprayer (litres)} \times 10\,000}{\text{Volume rate (litres/ha)} \times \text{Swathe width (metres)}}$

5. $\text{Volume of liquid herbicide product required when band spraying (litres)} = \dfrac{\text{Area of plantation (ha)} \times \text{Product rate (litres/ha)} \times \text{Swathe width (metres)}}{\text{Row spacing (metres)}}$

The area of plantation actually treated when band spraying is presented in ready reckoner form in Table 5.

6. $\text{Volume per spot (millilitres)} = \dfrac{\text{Volume rate (litres/ha)} \times \text{Area of spot (m}^2\text{)}}{10}$

7. $\text{Time per spot (seconds)} = \dfrac{\text{Required volume per spot (ml)}}{\text{Nozzle output} \div 60 \text{ (ml/min)}}$

8. $\text{Volume rate for spot treatment (litres)} = \dfrac{\text{Volume per spot (ml)} \times \text{Number of trees (spots)}}{1000}$

9. $\text{Number of trees sprayed with spot treatment} = \dfrac{\text{Volume of liquid in sprayer (litres)}}{\text{Volume per tree (litres)}}$

10. $\text{Distance covered by a tank full of spray when spot spraying (metres)} = \dfrac{\text{Volume of liquid in sprayer (litres)} \times 1000 \times \text{Tree spacing in rows (metres)}}{\text{Volume rate (litres/ha)} \times \text{Area of spot (m}^2\text{)}}$

11

11. **Quantity of herbicide product required (litres) per applicator/container**

$$= \frac{\text{Applicator/container capacity (litres)} \times \text{Product rate (litres/ha)}}{\text{Volume rate (litres/ha)}}$$

Table 4 The proportion of 1 hectare of plantation treated during spot application

Spot diameter (m)	Spot area (m²)	1.7	1.8	1.9	2.0	2.1	2.2	2.3	3.0
		Tree spacing (m) (assumes fully stocked square planting)							
1.5	1.77	.61	.55	.49	.44	.40	.37	.33	.19
1.4	1.54	.53	.48	.43	.39	.35	.32	.29	.17
1.3	1.33	.46	.41	.37	.33	.30	.27	.25	.14
1.2	1.13	.39	.35	.31	.28	.26	.23	.21	.13
1.1	.95	.33	.29	.26	.24	.22	.20	.18	.11
1.0	.79	.27	.24	.22	.20	.18	.16	.15	.09

Table 5 The proportion of 1 hectare of plantation treated during band application

Width of swathe (m)	1.7	1.8	1.9	2.0	2.1	2.2	2.3	3.0
				Row spacing (m)				
1.5	.88	.83	.79	.75	.71	.68	.65	.50
1.4	.82	.78	.74	.70	.67	.64	.61	.47
1.3	.76	.72	.68	.65	.62	.59	.57	.43
1.2	.71	.67	.63	.60	.57	.55	.52	.40
1.1	.65	.61	.58	.55	.52	.50	.48	.37
1.0	.59	.56	.53	.50	.48	.45	.43	.33

11

11.4 *Nozzles*

General information

With hydraulic nozzles, increasing the pressure will increase the nozzle output, reduce the size of the droplet and possibly widen the swathe width or spot diameter. Decreasing the pressure has the opposite effects.

All nozzles should be inspected regularly for wear and damage which can alter their outputs, droplet size and distribution characteristics.

Damage to the spinning discs of the rotary atomisers will affect droplet size.

All damaged or worn items should be replaced immediately.

Nozzle manufacturers have different conventions of defining pressures; bar, pounds per square inch (p.s.i.) and kilopascals (kPa). The units of conversion are:

$$1 \text{ bar} = 14.5 \text{ p.s.i.} = 100 \text{ kPa}$$

Hydraulic nozzles (fitted to the knapsack and Forestry Spot Gun)

The three types of hydraulic nozzles recommended in this Field Book are:

a. **Floodjet** (also known as anvil or deflector) **nozzles** produce a wide angled flat spray with an even distribution over the swathe width. At low pressures these nozzles produce large droplets with little risk of drift.

Flat even fan nozzles which also produce an even distribution over the swathe width are available in a range of angles and produce a similar pattern to the floodjet. These work at a slightly higher pressure thus producing smaller droplets with a higher risk of drift.

Normal flat fan nozzles are unacceptable for forestry as they do not produce an even distribution over the swathe width and therefore have to be overlapped to produce an even distribution over the treated area.

b. **Solid cone nozzles** distribute droplets over the entire area of a spot where the diameter is dependent on the angle of the nozzle and the height at which the nozzle is held.

c. **Solid stream nozzles** deliver the spray solution as a solid stream and are used for stem or stump treatments.

11

Feed nozzles (fitted to the Herbi's, ULVA and Ulvaforest)
Varying orifice sizes regulate the flow of the spray solution to the spinning disc.

For further information on nozzles refer to the *Nozzle selection handbook* published by the British Crop Protection Council, 20 Bridgport Road, Thornton Heath, Surrey CR4 7QC.

11.5 *Applicators for liquid herbicides*

11.5.1 Cooper Pegler CP15 forestry model knapsack sprayer with pressure control valve

Uses
Cut stump application at high volumes.
General herbicide application at medium and low volumes.

Supplier
Cooper Pegler Services Limited
Unit 5
Woodway Farm
Long Crendon
Bucks
HP18 6EP
Telephone: 0844 208703

11

Description
CP15 Forestry Model 15 litre knapsack sprayer, complete with pressure gauge and adaptor, pressure control valve and adaptor, and hose assembly. Ref. No. PA 1104.

Accessories
Spray shield 30 mm (12") complete with Green Polijet
Ref. No. SA 04 630
Spray shield 35 mm (15") complete with Green Polijet
Ref. No. SA 04 631
Politec Tree Guard with two nozzles. Ref. No. SA 04 637

Table 6 Nozzle-output data
a. Floodjet

Nozzle type	Nozzle output @ 0.7 bar (ml/min with water)	Nozzle output @ 1 bar (ml/min with water)
Polijet		
Red	2000	2400
Blue	1500	1800
Green (Note 1)	1000	1200
Very low volume (Notes 1, 2 and 3)		
AN 2.0 Blue	730	920
AN 1.0 Orange	360	460

Notes:
1. Also for use in the spray shield.
2. Although termed very low volume nozzles these apply at the low volume rate.
3. The Cooper Pegler brass VLV 100 and 200, which are no longer manufactured, equate to the AN 1.0 and AN 2.0 respectively with regards to nozzle output at 1.0 bar and litres per hectare.

b. Solid cone with green swirl core

Nozzle type	Nozzle output @ 0.7 bar (ml/min with water)	Nozzle output @ 1 bar (ml/min with water)
5 Blue	1000	1240
6 Yellow	1320	1670

11

Tools required for maintenance and calibration
Pliers
Large screwdriver
Medium screwdriver
Adjustable spanner
Roll of PTFE plumber's tape
Plastic bucket

Large funnel
Metric graduated measure
Clean rags or absorbent paper
Watch
1-metre length of 6 mm diameter metal rod or dowel

Calibration

Overall or band application

1. Determine the walking speed which an operator can sustain over the site with a full applicator and wearing the necessary protective clothing.
2. Select volume rate from the Method of application paragraph of the selected herbicide.
3. Enter Tables 7 and 8 by walking speed and volume rate to find the required nozzle output (ml/min) over a 1 metre swathe to achieve an even distribution for these two parameters.

Table 7 Required nozzle output at low volume (ml/min) for
            1 metre swathe

Sustainable walking speed for site (m/min)	Low volume 50–200 l/ha						
	50	75	100	125	150	175	200
60	300	450	600	750	900	1050	1200
55		413	550	688	825	963	1100
50		375	500	625	750	875	1000
45		338	450	563	675	788	900
40		300	400	500	600	700	800
35			350	438	525	613	700
30			300	375	450	525	600
25			250	313	375	438	500
20			200	250	300	350	400

11

Table 8 Required nozzle output at medium volume (ml/min) for
1 metre swathe

Sustainable walking speed for site (m/min)	Medium volume 200–400 l/ha							
	225	250	275	300	325	350	375	400
60	1335	1500	1650	1800	1950	2100	2250	2400
55	1238	1375	1513	1650	1788	1925	2063	2200
50	1125	1250	1375	1500	1625	1750	1875	2000
45	1013	1125	1238	1350	1463	1575	1688	1800
40	900	1000	1100	1200	1300	1400	1500	1600
35	788	875	963	1050	1138	1225	1313	1400
30	675	750	825	900	975	1050	1125	1200
25	563	625	688	750	813	875	938	1000
20	450	500	550	600	650	700	750	800

Notes (for Tables 7 and 8):
i. Interpolate where necessary.
ii. Nozzle outputs calculated from equation:

$$\text{Nozzle output (ml/min)} = \frac{\dfrac{\text{Walking speed}}{\text{(metre/minute)}} \times \dfrac{\text{Volume rate}}{\text{(litre/hectare)}} \times \dfrac{\text{Swathe width}}{\text{(metres)}}}{10}$$

iii. Use the above equation to calculate nozzle outputs between 400–700 l/ha.

4. To calculate nozzle output (ml/min) for a swathe wider
    than one metre multiply the nozzle output (given in
    Tables 7 and 8) by the required swathe width.
    Example:
    Walking speed   45 m/min
    Volume rate   200 l/ha
    Swathe width   1.2 m
    Required nozzle output from low volume table
                                        =900 ml/min
    Required nozzle output for 1.2 m swathe   =900 × 1.2
                                        =1080 ml/min
5. From the nozzle output data select and fit to the knap-
    sack the nozzle which gives the nearest (but greater)
    output than the required nozzle output. The green
    polijet would be the nozzle selected for the above
    example.
6. Fill the applicator with water and fully pressurise.

11

7. Set the pressure gauge by adjusting the screw of the pressure control valve to an initial pressure of 1.0 bar.

8. Spray into a graduated measure for a set time and calculate nozzle output ml/min.

9. Compare the actual nozzle output with the required nozzle output. If there is a difference adjust the pressure and re-measure output until the actual and required nozzle outputs agree.

10. When outputs agree note the pressure gauge reading for that sprayer. Pressure should be checked two or three times during the day using the same pressure gauge.

11. Having filled the applicator in the approved manner with herbicide diluted to the required volume, check nozzle output. If different re-calibrate.

Spot application
1. Select volume rate from Method of application paragraph of the selected herbicide.

2. Decide on spot diameter required.

3. Enter Tables 9 and 10 by volume rate and spot diameter.

Example
Volume rate   175 l/ha
Spot diameter   1.2 m
Volume per spot=19.8 ml

4. From the nozzle output data select and fit to the knapsack the nozzle which gives the nearest (but greater) output than the required nozzle output.

5. Fill the applicator with water and fully pressurise.

6. Set the pressure gauge by adjusting the screw of the pressure control valve to an initial pressure of 1 bar.

7. Spray into a graduated measure for a set time to obtain nozzle output in ml/min.

8. To calculate the application time required per spot in seconds, divide the required volume per spot (from Tables 9 and 10) by the result of: $\frac{\text{nozzle output (ml/min)}}{60}$.

9. Practice application into a graduated measure until the desired volume per spot is achieved; then over a dry

Table 9 Volume per spot (ml)

Spot diameter (metres)	Low volume 50–200 l/ha						
	50	75	100	125	150	175	200
1.0			7.9	9.8	11.8	13.8	15.8
1.1		7.1	9.5	11.9	14.3	16.6	19.0
1.2		8.5	11.3	14.1	17.0	19.8	22.6
1.3	6.7	10.0	13.3	16.6	19.9	23.3	26.5
1.4	7.6	11.5	15.4	19.2	23.1	26.6	30.8

Table 10 Volume per spot (ml)

Spot diameter (metres)	Medium volume 200–400 l/ha							
	225	250	275	300	325	350	375	400
1.0	17.7	19.6	21.6	23.6	25.5	27.5	29.5	31.4
1.1	21.4	23.8	26.1	28.5	30.9	33.3	35.6	38.0
1.2	25.4	28.3	31.1	33.9	36.7	39.6		
1.3	29.9	33.2	36.5	39.8				
1.4	34.7	38.4						

surface to achieve the required spot size. If the application time is not practical the volume rate should be adjusted until a realistic time is achieved.

10. Having filled the applicator in the approved manner with herbicide diluted to the required volume, check nozzle output. If different re-calibrate.

Droplet size
If undesirable drift is occurring use a larger droplet size by changing to a nozzle of the same type but with a larger hole and use at a lower pressure to obtain the same nozzle output. If necessary, adjust the working height of the nozzle to obtain the same width of treatment.

Cleaning
Spray out all dilute pesticide safely (see Section 3.4). Wash thoroughly with a weak solution of detergent or washing soda and water. Shake well and spray out after removing the nozzle. Finally, pump through clean water.

11.5.2   Tecnoma T16P knapsack sprayer

Use
Ammonium sulphamate application to cut stump at high volume and following frill girdling.

Supplier
Tecnoma Sprayers Limited
Blue House Lane
Mendlesham
Stowmarket
Suffolk
IP14 5NH
Telephone: 0449 767757

Description
Tecnoma T16P; 16 litre knapsack sprayer fitted with a stainless steel ball bearing in the pump which will not be corroded by ammonium sulphamate. The return spring in the trigger assembly will corrode and will require replacement.

Tools required for maintenance
No tools are required as the sprayer can be dismantled by hand. Plastic bucket and clean rags or absorbent paper should be available.

Calibration
As the sprayer has no pressure controls and as ammonium sulphamate is applied to the point of run off, no calibration is necessary.

11

Cleaning
Spray out all dilute pesticide safely (see Section 3.4). Wash and flush through with a weak solution of warm detergent and rinse with clean water several times.

11.5.3   Forestry Spot Gun

Uses
Cut stump application at high volume.
Woody weed foliar application at low volume.
Grasses and herbaceous broadleaved application at low and very low volume.

Supplier
Selectokil
Abbey Gate Place
Tovil
Maidstone
Kent
ME15 0PP
Telephone: 0622 55471

Description
Forestry Spot Gun; adapted from the veterinary drench
gun with a 5 litre knapsack. This applicator can be
adjusted to give a measured dose up to 20 ml.

Accessories
Spare knapsack from Selectokil.
Steel cylinder for use with certain pesticides from
Selectokil.
Spare nozzles from Selectokil.
25 litre Rigid polythene bottle with 18 mm tap bore.
Ref. BA/PS4
From: Fisons Scientific Equipment Division
　　　　Bishops Meadow Road
　　　　Loughborough
　　　　Leicestershire
　　　　LE11 0RG

Nozzle data

Nozzle type	For treating
TG 2.8 W wide angle solid cone	Grass/herbaceous broadleaved weeds
TG5 narrow angle solid cone	Woody weeds
No. 0006 or LF6-0 solid stream	Chemical thinning or cut stump

Tools required for maintenance and calibration
Pliers
Two 25 mm adjustable wrenches

Funnel
Plastic bucket
Metric graduated measure
Clean rags or absorbent paper
Tape measure
Lubricating oil
Roll of PTFE plumber's tape

Calibration

1. Adjust gun to deliver a 5 ml dose.
2. Fill knapsack with 5 litres of water and mark this level (filler cap upmost).
3. Prime gun.
4. With the nozzle held at spray height over a dry surface, squeeze the trigger a number of times until a well defined spot appears. Measure spot diameter, disregarding peripheral droplets.
5. Enter Table 11 below by spot diameter and application rate of herbicide product to determine the quantity of product required per 5 litre knapsack.

Example:
Spot diameter 1.2 m.
Rate of application for glyphosate to control grasses in lowland Britain (Section 4.5) 1.5 litres per hectare.
  Therefore from Table 11 the quantity of herbicide required per 5 litres is 0.17 litres (i.e. 0.17 litres of herbicide + 4.83 litres of water = 5 litres).

6. Fill knapsack in the approved manner with herbicide/ diluent mixture.
7. Check the spot diameter as in paragraph 4; if different re-calibrate.

11

Cleaning

Spray out all dilute pesticide safely (see Section 3.4). Wash and flush through with a weak solution of warm water and detergent, rinse with clean water. Note damage to the applicator can occur if subject to temperature in excess of 50°C.
  To lubricate split the gun by undoing the handle cramp screw, remove the complete piston assembly and apply a liberal quantity of oil to the moving parts.

Table 11 Product rate (herbicide product litres per hectare)

Spot diameter (metres)	1.0	1.5	2.0	3.0	3.75	4.0	5.0	6.0	7.0	8.0	9.0	10.0	11.0	12.0	13.0	13.5	14.0	15.0
0.60	0.03	0.04	0.06	0.08	0.11	0.11	0.14	0.17	0.20	0.23	0.25	0.28	0.31	0.34	0.37	0.38	0.40	0.42
0.65	0.03	0.05	0.07	0.10	0.12	0.13	0.17	0.20	0.23	0.27	0.30	0.33	0.37	0.40	0.43	0.45	0.46	0.50
0.70	0.04	0.06	0.08	0.12	0.14	0.15	0.19	0.23	0.27	0.31	0.35	0.38	0.42	0.46	0.50	0.52	0.54	0.58
0.75	0.04	0.07	0.09	0.13	0.17	0.18	0.22	0.27	0.31	0.35	0.40	0.44	0.49	0.53	0.57	0.60	0.62	0.66
0.80	0.05	0.08	0.10	0.15	0.19	0.20	0.25	0.30	0.35	0.40	0.45	0.50	0.55	0.60	0.65	0.68	0.70	0.75
0.85	0.06	0.09	0.11	0.17	0.21	0.23	0.28	0.34	0.40	0.45	0.51	0.57	0.62	0.68	0.74	0.77	0.79	0.85
0.90	0.06	0.10	0.13	0.19	0.24	0.25	0.32	0.38	0.45	0.51	0.57	0.64	0.70	0.76	0.83	0.86	0.89	0.95
0.95	0.07	0.11	0.14	0.21	0.27	0.28	0.35	0.43	0.50	0.57	0.64	0.71	0.78	0.85	0.92	0.96	0.99	1.06
1.00	0.08	0.12	0.16	0.24	0.29	0.31	0.39	0.47	0.55	0.63	0.71	0.79	0.86	0.94	1.02	1.06	1.10	1.18
1.05	0.09	0.13	0.17	0.26	0.32	0.35	0.43	0.52	0.61	0.69	0.78	0.87	0.95	1.04	1.13	1.17	1.21	1.30
1.10	0.10	0.14	0.19	0.29	0.36	0.38	0.48	0.57	0.67	0.76	0.86	0.95	1.05	1.14	1.24	1.28	1.33	1.43
1.15	0.10	0.16	0.21	0.31	0.39	0.42	0.52	0.62	0.73	0.83	0.93	1.04	1.14	1.25	1.35	1.40	1.45	1.56
1.20	0.11	0.17	0.23	0.34	0.42	0.45	0.57	0.68	0.79	0.90	1.02	1.13	1.24	1.36	1.47	1.53	1.58	1.70
1.25	0.12	0.18	0.25	0.37	0.46	0.49	0.61	0.74	0.86	0.98	1.10	1.23	1.35	1.47	1.60	1.66	1.72	1.84
1.30	0.13	0.20	0.27	0.40	0.50	0.53	0.66	0.80	0.93	1.06	1.19	1.33	1.46	1.59	1.73	1.79	1.86	1.99
1.35	0.14	0.21	0.29	0.43	0.54	0.57	0.72	0.86	1.00	1.15	1.29	1.43	1.57	1.72	1.86	1.93	2.00	2.15
1.40	0.15	0.23	0.31	0.46	0.58	0.62	0.77	0.92	1.08	1.23	1.39	1.54	1.69	1.85	2.00	2.08	2.16	2.31
1.45	0.17	0.25	0.33	0.50	0.62	0.66	0.83	0.99	1.16	1.32	1.49	1.65	1.82	1.98	2.15	2.23	2.31	2.48
1.50	0.18	0.27	0.35	0.53	0.66	0.71	0.88	1.06	1.24	1.41	1.59	1.77	1.94	2.12	2.30	2.39	2.47	2.65
1.55	0.19	0.28	0.38	0.57	0.71	0.75	0.94	1.13	1.32	1.51	1.70	1.89	2.08	2.26	2.45	2.55	2.64	2.83
1.60	0.20	0.30	0.40	0.60	0.75	0.80	1.00	1.21	1.41	1.61	1.81	2.01	2.21	2.41	2.61	2.71	2.81	3.02

11.5.4   **Microfit Herbi**

Use
Band or overall application at very low volumes.

Supplier
Controlled Droplet Application Limited
Lockinge
Wantage
Oxfordshire
OX12 8PH
Telephone: 0235 833270

Description
Microfit Herbi a low speed rotary atomiser producing
droplets with a volume median diameter (VMD) of 250
microns.

Accessories
2.5 litre plastic bottles from CDA Ltd.
Spare motor or atomiser heads from CDA Ltd.
Battery tester with voltage ranges of 0–1.5 v and 0–12 v,
Panda Multimeter Model ELT 801 (obtain from a local
electrical retailer).
Alkaline manganese batteries (obtain locally).

Tools required for maintenance and calibration
Pliers
Philips screwdriver
Plastic bucket
Large funnel
Metric graduated measure
Clean rags or absorbent paper
Watch
50 m tape measure

Calibration
1.   Determine the walking speed which an operator can
     maintain over the site with a full applicator and wear-
     ing the necessary protective clothing.
2.   Select volume rate from the Method of application para-
     graph of the selected herbicide.
3.   Enter Table 12 by walking speed and volume rate to
     find the required nozzle output (ml/min) over a 1.2
     metre swathe.

11

Table 12 Required nozzle output for 1.2 metre swathe (ml/min)

Sustainable walking speed for site (m/min)	Volume rate l/ha											
	10.0	12.0	14.0	16.0	18.0	20.0	22.0	24.0	26.0	28.0	30.0	32.0
60	72	87	101	116								
55	66	79	93	106	119							
50	60	72	84	96	108	120						
45		65	76	85	98	108	119					
40		58	67	77	87	96	106	116				
35			59	67	76	84	93	101	110	118		
30				58	65	72	80	87	94	101	108	116
25						60	66	72	78	84	90	96
20								58	63	68	72	77
15												58

Note: Interpolate where necessary.

4. The ideal feed rate to the atomisers is 60 ml/min. At 120 ml/min and above, atomisation is not so efficient; small satellite droplets may form and present a drift hazard.

5. *Always check swathe width.*
   For a swathe wider than 1.2 metres the calculated nozzle output (from Table 12) has to be adjusted by the following equation:

$$\text{Required nozzle output for new swathe width} = \frac{\text{Required nozzle output for 1.2 m swathe (ml/min)} \times \text{Actual swathe width (m)}}{1.2}$$

Example:
Walking speed   40 m/min
Volume rate   20 l/ha
Actual swathe width 1.4 m

From Table 12 the required nozzle output for a 1.2 metre swathe is 96 ml/min. Therefore the required nozzle output for a 1.4 metre swathe width is:

$$\frac{96 \times 1.4}{1.2} = 112 \text{ ml/min.}$$

6. Remove the spinning disc from the applicator.
7. Fit the feed nozzle which has the nearest nozzle output to that calculated.

Nozzle currently available with Microfit Herbi

Nozzle colour	Nominal nozzle output with water
Blue	90 ml/min
Yellow	150 ml/min
Orange	240 ml/min
Red	360 ml/min

Herbi 77 (no longer manufactured).

Nozzle colour	Nominal nozzle output with water
Blue	66 ml/min
Yellow	120 ml/min
Red	330 ml/min

Note: Viscosity of the liquid will alter the flow rates when using nozzle with Microfit Herbi or Herbi 77.

8. Attach a bottle containing the mixture of herbicide to be sprayed to the applicator.
9. Hold the applicator at an angle of about 40 degrees. Before placing the nozzle over a graduated measure ensure that all air bubbles are removed from the nozzle in order to obtain a constant flow. Measure the nozzle output for a set time to obtain nozzle output in ml/min.
10. Check that the measured output agrees with that calculated in paragraph 3 or 5. If they do not agree recalibrate by either:

    a. if the outputs are close, adjust the angle of the applicator. Increasing the angle up to 10 degrees will increase the flow and reducing the angle up to 10 degrees will reduce the flow. If this fails to give the required output, fit a different nozzle of the same colour.
    Note: There can be considerable variation in nozzle output within the same colour nozzles.

11

or b. if there is a considerable difference in output, fit a nozzle of a different colour;

or c. adjust volume rate if (a) and (b) do not achieve the required nozzle output;

or d. if (b) and (c) fail to achieve the required nozzle output adjust walking speed.

11. Check swathe width and re-calibrate if necessary.

12. Refit spinning disc.

Droplet size

Approximately 250 micron droplet VMD with the spinning disc revolving at 2000 rpm providing the voltage is between 4 and 6 volts. Check the voltage at the plug between the battery carrier and lance assembly; if below 4 volts replace batteries.

Herbi 77 is no longer manufactured but many are still in use; treat as Microfit Herbi except for voltage range. The voltage range for this applicator is 4 to 12 volts. Check the voltage at the two electrical terminals on the outside of the atomiser head; if below 4 volts replace batteries.

Cleaning

Spray out all dilute pesticide safely (see Section 3.4).

Remove batteries and clean off any corrosion on the contact points before storage. Wash thoroughly with a weak solution of detergent and rinse with clean water.

11

11.5.5 **ULVA 8**

Use
Incremental application

Supplier
Controlled Droplet Application Limited
Lockinge
Wantage
Oxfordshire
OX12 8PH
Telephone: 0235 833270

Description
High speed (5500—8800 rpm) rotary atomisers producing driftable droplets with a volume median diameter (VMD) of 70 microns.

ULVA 8 powered by 8 batteries. Its use is for incremental spraying in crops up to an average height of 1.5 m and for inter-row spraying of heather.

Accessories
Spare 1 litre plastic bottles from CDA Ltd.

Vibrotac high range rev counter (one per gang) from CDA Ltd.

Spare motor, spinning disc or complete head assembly from CDA Ltd.

Batteries: alkaline manganese batteries are recommended. If these batteries are used for 4 hours, then rested for about 20 hours before being used again they will give 25—30 hours total spraying time. Purchase locally. For larger programmes, rechargeable cadmium nickel batteries can be used. However, approximately 80 spraying days are required to break-even in cost with the alkaline manganese batteries from CDA Ltd.

Dwyer Wind Meter from CDA Ltd.

Smoke powder: minimum purchase 10×125 g tins.
From: Brocks Fireworks Limited
        Galeside
        Sanguhar
        Dumfries
        Telephone: 06592 531
Spare feed nozzles from CDA Ltd.

11

Tools required for maintenance and calibration
Pliers
Screwdrivers $2\frac{1}{2}''$ and $3''$
Screwdriver No. 2 Pozi 5332
Small adjustable wrench
Roll of adhesive tape
Plastic bucket
Large funnel
Metric graduated measure
Clean rags or absorbent paper
Watch
50 m tape measure

Calibration

Incremental application

1. Determine the walking speed which an operator can sustain over the site with a full applicator and wearing the necessary protective clothing.

2. Select volume rate from the Method of application paragraph of the selected herbicide.

3. Enter Table 13 by walking speed and volume rate to find the required nozzle output (ml/min) over a 5 metre swathe to achieve an even distribution for these two parameters. If these two parameters are outside the nozzle output range of the table, select an alternative volume rate.

4. Although the ideal nozzle output is 60 ml/min there are some circumstances where, in order to avoid walking at an extremely slow speed, the flow rate can be allowed to rise to 120 ml/min or even 180 ml/min, even though this will result in larger, rather than more droplets being formed.

Table 13 Required nozzle output for 5 metre swathe (ml/min)

Sustainable walking speed for site (m/min)	Volume rate l/ha					
	10.0	12.0	14.0	16.0	18.0	20.0
36	180					
34	170					
32	160					
30	150	180				
28	140	168				
26	130	156	182			
24	120	144	168			
22	110	132	154	176		
20	100	120	140	160	180	
18	90	108	126	144	162	180
16	80	96	112	128	144	160
14	70	84	98	112	126	140
12	60	72	84	96	108	120
10	50	60	70	90	90	100

11

5. If because of distances between rows a different swathe width is required within the operating range of 4–6 metres, the required nozzle output (from Table 13) has to be adjusted by the following equation:

Required nozzle   Required nozzle   Actual swathe
output for new   =   output for 5 m ×   width×0.2
swathe width   swathe (ml/min)   (m)
(ml/min)

Example:
Walking speed   26 m/min
Volume rate   12 l/ha
Actual swathe width   4.7 m
   From Table 13 the required nozzle output for a 5 metre swathe is 156 ml/min. Therefore the required nozzle output for a 4.7 metre swathe width is:

$$156 \times 4.7 \times 0.2 = 147 \text{ ml/min.}$$

6. Remove the spinning disc from Ulva.

7. Fit the feed nozzle which has the nearest nozzle output to that calculated.

Nozzle colour	Nominal nozzle output with thin oil
Yellow	30 ml/min
Red	60 ml/min
Grey	120 ml/min
Green	180 ml/min

Note: Viscosity of the liquid will alter the flow rates.

11

8. Attach a bottle containing the mixture of the herbicide to be sprayed to the applicator.

9. Before placing the nozzle over a graduated measure ensure that all air bubbles are removed from the nozzle in order to obtain a constant flow. Measure the nozzle output for a set time to obtain nozzle output in ml/min.

10. Check that the measured output agrees with that calculated in paragraphs 3 or 5. If they do not agree recalibrate by either:

   a. if the measured and required outputs are close, fit a different nozzle of the same colour;

Note: There can be considerable variation in nozzle output within the same colour nozzles.

or b. if there is a considerable difference in output fit a nozzle of a different colour;

or c. adjust volume rate if (a) and (b) do not achieve the required nozzle output;

or d. if (b) and (c) fail to achieve the required nozzle output adjust walking speed.

11. If during the day the temperature changes, the nozzle output must be checked and if necessary re-calibrated.

Directed application—heather control

1. Determine the walking speed which an operator, whilst sweeping the applicator from side to side, can sustain over the site with a full applicator and wearing the necessary protective clothing.

2. Select volume rate from the Method of application paragraph of the selected herbicide.

3. Enter Table 14 by walking speed and volume rate to find required nozzle output (ml/min) over a 2 metre swathe.

Table 14 Required nozzle output for 2 metre swathe (ml/min)

| Sustainable walking speed for site (m/min) | Volume rate l/ha | | | |
	10.0	12.0	15.0	16.0
40	80	96	120	128
38	76	91	114	122
36	72	87	108	116
34	68	82	102	109
32	64	77	96	103
30	60	72	90	96
28	56	68	84	89
26	52	63	78	84
24	48	58	72	77
22	44	53	66	71
20	40	48	60	64
18	36	44	54	58
16	32	39	48	52
14	28	34	42	45
12	24	29	36	39
10	20	24	30	32

4. Although the ideal nozzle output is 60 ml/min there are some circumstances where, in order to avoid walking at an extremely slow speed, the flow rate can be allowed to rise to 120 ml/min or even 180 ml/min, even though this will result in larger, rather than more droplets being formed.

5. If a different swathe width is required the nozzle output (from Table 14) has to be adjusted by the following equation:

$$\text{Required nozzle output for new swathe width} = \frac{\text{Required nozzle output for 2 m swathe (ml/min)} \times \text{Actual swathe width (m)}}{2}$$

Example:

Walking speed	18 m/min
Volume rate	12 l/ha
Actual swathe width	1.75 m

From the table the required nozzle output for a 2 metre swathe is 44 ml/min.
Therefore the required nozzle output for a 1.75 metre swathe width is:

$$\frac{44 \times 1.75}{2} = 38.5 \text{ ml/min.}$$

6. Remove the spinning disc from Ulva.

7. Fit the feed nozzle which has the nearest nozzle output to that calculated.

Nozzle colour	Nominal nozzle output with thin oil
Yellow	30 ml/min
Red	60 ml/min
Grey	120 ml/min
Green	180 ml/min

Note: Viscosity of the liquid will alter the flow rates.

8. Attach a bottle containing the mixture of the herbicide to be sprayed to the applicator.

9. Before placing the nozzle over a graduated measure ensure that all air bubbles are removed from the nozzle

in order to obtain a constant flow. Measure the nozzle output for a set time to obtain nozzle output in ml/min.

10. Check that the measured output agrees with that calculated in paragraph 3 or 5. If they do not agree recalibrate by either:

    a. if the measured and required outputs are close, fit a different nozzle of the same colour.
       Note: There can be considerable differences in nozzle output within the same colour nozzles.

or b. if there is a considerable difference in output, fit a nozzle of a different colour;

or c. adjust volume rate if (a) and (b) do not achieve the required nozzle output;

or d. if (b) and (c) fail to achieve the required nozzle output adjust walking speed.

11. Refit spinning disc.

12. If during the day the temperature changes, the nozzle output must be checked and if necessary re-calibrated.

Droplet size
The droplet size is approximately 70 microns VMD when the disc speed is between 5500—8000 rpm. Disc speed should be checked regularly with the 'Vibrotac'; if below 5500 RPM the batteries should be changed.

Wind speed
Wind speed should always be measured at the height of the sprayer.

For incremental application over an approximate 5 metre swathe, a minimal speed of 3 km per hour (2 mph) is acceptable with a maximum speed of 12 km per hour (7 mph), occasionally gusting to 16 km per hour (10 mph).

For directed application on heather 0—19 km per hour (0—12 mph) is suitable.

Wind direction
Incremental application can be carried out if the angle between the operator's line of walking and the wind direction is not less than 20 degrees.

Safety zone
A 100 metre zone, downwind, should be allowed where there are susceptible crops, especially those on adjoining

11

farmland and crops growing in polytunnels or under glass. Do not spray if the wind is blowing towards susceptible glass house and similar crops.

Cleaning
Remove batteries and clean off any corrosion on the contact points before storage.

Wash thoroughly with a weak solution of detergent and rinse with clean water.

11.5.6 **Stihl SG17 mistblower**

Use
Herbicide application at low volumes.

Supplier
Andreas Stihl Limited
Goldsworth Park Trading Estate
Woking
Surrey
GU21 3BA
Telephone: 04862 20222

Description
Motorised knapsack mistblower with a tank capacity of 11 litres. Infinitely variable rate of discharge between 0.8 and 4.0 l/min. Weight when full approximately 20 kg.

Maintenance and calibration
Details of maintenance and calibration can be obtained from:
Work Study Branch
Forestry Commission
Santon Downham
Brandon
Suffolk
IP27 0TJ
Telephone 0842 814227

11

11.5.7 **Handheld direct applicator**
Weedwiper Mini.

Supplier
Hortichem Limited
14 Edison Road
Churchfield Industrial Estate
Salisbury
Wiltshire
Telephone: 0722 20133

Description
SM90 Weedwiper Mini Standard Single-head (Standard
Blue Label Wick) fitted with venting cap.
 SMR90 Weedwiper Mini Red Band Single-head (Fast
Flow Red Label Wick) fitted with venting cap.
 The SM90 fitted with the standard wick is recommended
for use on normal vegetation. Where the vegetation is
dense or semi-resistant, the SMR90 with the fast flow wick
should be used.

Accessories
Spare wicks from: Hortichem Limited.
10×55 g sachets of red dye from Hortichem Limited.
5-litre (Ref. PO 23/01) or 10 litre (Ref. PO 23/02) graduated
polythene bottle with screw cap, tap and metal carrying
handle.
From: Fisons Scientific Equipment Division
      Bishop Meadow Road
      Loughborough
      Leicestershire
      LE11 0RG

Tools required for maintenance and calibration
Small adjustable wrench
Small funnel to fit filler cap
Metric graduated measure
Clean rags or absorbent paper
Plastic bucket
Teaspoon
Scrubbing brush
Plastic bags

Calibration

As it is not possible to calibrate the weedwiper, the only way of ensuring that the wick is wet enough to transfer herbicide on to the vegetation is by adjusting the flow during use.

To adjust flow:

1. Mark the reservoir (handle) at 75 cm from the filler cap. Always keep the level of herbicide above this point during use to ensure adequate wick flow.
2. Fill applicator with herbicide mixture to which has been added 1 teaspoonful of red dye per litre.
3. Open venting cap.
4. Allow wick to become impregnated with herbicide mixture. When fully impregnated the wick should be red.
5. Start treating the vegetation (see Working method).
6. If the wick blanches during use, the compression nuts should be released to increase wick flow. It may also be necessary to pull the rubber seals from their seats and re-adjust their position on the wick.
7. If the wick remains bright red and drips during use, tighten the compression nuts to reduce wick flow.

Working method

A back and forth sweeping action, taking care not to touch the crop tree, will ensure that both sides of the weeds are treated. Where the trees are standing above the weeds, application should be made in a triangular pattern. A triple width pass on each side of the triangle will give control over a 1.08 m diameter spot. For trees below weed level, use 1.0 m long, back and forth passes over the tree, ensuring there is a 10 cm clearance between the wick and the top of the tree, to give control over 0.90 square metre.

11

Cleaning

In order to achieve maximum transfer of herbicide on to the weeds, any dirt adhering to the wick should be scrubbed off immediately. It is recommended that each day after use the applicator head is either immersed in water or the applicator filled with water and left soaking overnight to cleanse the wick.

Wicks should be replaced regularly.

New wicks should be fitted after storage or if the old wicks have dried out.

Wash thoroughly with a weak solution of detergent and rinse the weedwiper with clean water.

11.5.8   ULVAFOREST tractor-mounted sprayer

Use

Tractor-mounted sprayer for band or complete application at very low volume.

Supplier

Controlled Droplet Application Ltd
Lockinge
Wantage
Oxfordshire
OX12 8PH
Telephone: 0235 833270

Note: These sprayers are only built to order.

Description

The Ulvaforest consists of a 270 litre tank and a fully gimballed 4.8 metre boom mounted on to a hydraulic ram which allows it to be raised to 2 metres. Five Microfit Herbi rotary atomisers are fitted to the boom, each being controlled by an individual switch in the tractor cab. There are three basic spraying patterns, i.e. overall with a 6 metre swathe or over-the-rows spraying of either two or three rows with a nominal 1.2 metre swathe.

Maintenance and calibration

Details of maintenance and calibration can be obtained from:

Work Study Branch
Forestry Commission
Santon Downham
Brandon
Suffolk
IP27 0TJ
Telephone: 0842 814227

11.6 *Applicators for spot application of granules*

11.6.1 **Pepperpots**
These are supplied free with Atlas Lignum and propyzamide granules.

Accessories
A belt to hold 13 propyzamide beaker-type pots can be
obtained from:
Kingswood Canvas Ltd
197 Two Mile Hill Road
Kingswood
Bristol
BS15 1AZ
  A haversack to hold the Atlas Lignum pepperpots
should be purchased locally.
  For transportation the lids of the pepperpots should be
sealed with adhesive tape. Exchange the sealed lid with an
unsealed one for application.

*Application technique and calibration (see Section
11.6.2).*

11.6.2 **Moderne applicator**
Use without the dose rate regulator fitted, from:
Stewart and Co.
Stronghold Works
Mayfield Industrial Estate
Dalkeith
Midlothian
EH22 4BZ

Application technique
For both the pepperpot and the Moderne, the technique is
to shake the applicator from side to side whilst traversing
the spot.

11

Calibration
1. Rate of application.

Application rate (product per hectare)

Per hectare		Per m²
30.0 kg/ha	=	3.0 g/m²
38.0 kg/ha	=	3.8 g/m²
40.0 kg/ha	=	4.0 g/m²

2. To calculate the quantity of granules required to treat a spot greater or lesser than one square metre, multiply the actual spot area to be treated by the quantity of product needed to treat 1 m².

Example:
Quantity of granules required to treat a 1.2 m² spot at 38.0 kg/ha is:

$$3.8 \text{ g} \times 1.2 \text{ m} = 4.56 \text{ g per spot}$$

Applicator calibration
1. Fill applicator with granules and weigh.
2. Using the side to side application technique shake the applicator for a set number of times over a paper or plastic sheet.
3. Measure the spot area and weigh the applicator or remaining granules.
4. Calculate whether the amount of granules applied is correct for the area treated (i.e. spot area × number of spots treated).
5. If not correct adjust the vigour of shake and number of passes over the spot and re-calculate.

Cleaning
Wipe the inside and outside of the applicator with a dry cloth or paper towel.

11

11.7 Tree injection applicators

11.7.1 Jim Gem

Use
Stem injection of woody weeds.

Supplier
Stanton Hope Limited
11 Seax Court
Southfields
Laindon
Basildon
Essex
SS15 6LY
Telephone: 0268 419141

Description
A 1.2 metre long metal tube, which acts as the reservoir, with a chisel bit fitted at one end. A lever operates the pump which can be adjusted to deliver between 0.5 and 2.0 ml through the chisel into the cut.

11.7.2 Hypo Hatchet

Use
Stem injection of woody weeds.

Supplier
Stanton Hope Limited
11 Seax Court
Southfields
Laindon
Basildon
Essex
SS15 6LY
Telephone: 0268 419141

Description
A small hatchet which is bored to allow herbicide to be fed from a reservoir strapped around the operator's waist. When the hatchet hits the tree a 1 ml dose of herbicide is injected.

11

11.8   *ENSO brush cutter stump treatment attachment*

Use
Application of herbicide directly on to cut stumps during cutting operation.

Supplier
Chieftain Forge
Burnside Road
Bathgate
West Lothian
EH48 4PH
Telephone: 0506 52354
   When ordering, detail make and model number of brush cutter.

Description
The kit consists of a supply tank, pump operated by a twist grip, nozzle and pipework for fitting to the brush-cutter. In use, by twisting the grip and activating the pump, herbicide is sprayed out on to the underside of the cutting blade thus depositing herbicide on to the freshly cut stump whilst cutting.

Maintenance and calibration
Details of maintenance and calibration can be obtained from:
Work Study Branch
Forestry Commission
Santon Downham
Brandon
Suffolk
IP27 0TJ
Telephone: 0842 814227

11

11.9 *Pesticide transit box*

Box suitable for transporting pesticides in vehicles.

Internal dimensions
Length   53 cm
Width   37 cm
Height   35 cm

Supplier
Cairn Craft
The Factory
Fogers (by Loch Ness)
Inverness
Telephone: 04563 285

11

12 List of herbicides and adjuvants etc and their manufacturers or distributors

Herbicide	Manufacturer's or distributor's code number (see following list)
A Agral	8
B Amcide	2
C Asulox	6
D Atlas Lignum	1
E Atrazine	4,11
F 2, 4-D ester	3,11
G Gardoprim	4
H Holtox	11
I Hortichem scarlet dye	7
J Kerb	10
K Mixture B	9
L Pbi spreader	10
M Roundup	9
N Timbrel	5
O Velpar	12

12

Manufacturers or distributors of approved products

		Herbicide code letters
1.	Atlas Interlates Ltd. Fraser Road Erith Kent DA8 1PN 032 24 32255	D
2.	Battle, Hayward & Bower Ltd. Victoria Chemical Works Crofton Drive Allenby Road Industrial Estate Lincoln LN3 4NP 0522 29206/41241	B
3.	BP Oil Ltd. Crop Protection Dept. BP House Breakspears Way Hemel Hemstead Herts HP2 4UL 0442 225787	F
4.	Ciba-Geigy Agrochemicals Whittlesford Cambridge CB2 4QT 0223 833621	E, G
5.	Dow Agriculture Latchmore Court Brand Street Hitchin SG5 1HZ 0462 57272	N

12

6. Embetec Crop Protection   C
   Springfield House
   Kings Road
   Harrogate
   HG1 5JJ
   0423 509731/5

7. Hortichem Ltd.   I
   1 Edison Rd
   Churchfields Industrial Estate
   Salisbury
   Wiltshire
   SP2 7NU
   0722 20133

8. ICI Agrochemicals   A
   Woolmead House West
   Bear Lane
   Farnham
   Surrey
   GU9 7UB
   0252 733888

9. Monsanto Ltd.   K, M
   Agricultural Division
   Thames Tower
   Burleys Way
   Leicester
   LE1 3TP
   0533 20864

10. Pan Britannica Industries Ltd.   J, L
    Britannica House
    Waltham Cross
    Herts
    EN8 7DY
    0992 23691

11. Rhone-Poulenc Environmental Products   E, F, G
    Regent House
    Hubert Road
    Brentwood
    Essex
    CM14 4TZ
    0277 261414

12

12. Selectokil Ltd.   O
     Abbey Gate Place
     Tovil
     Maidstone
     Kent
     ME15 OPP
     0622 55471

12

13 Forestry Commission sources of advice

Conservancy and Forest District Offices
Policy and general operational advice
(see telephone directory for local offices)

Research Division
Silviculturist (South)
Herbicides in lowland forests
Alice Holt Lodge, Wrecclesham, Farnham, Surrey
GU10 4LH
0420 22255

Silviculturist (North)
Herbicides in upland forests
Northern Research Station, Roslin, Midlothian EH25 9SY
031—445 2176

Work Study Officer
Protective clothing and appliances
Eastern Region Work Study Team, c/o Forestry
Commission District Office, Santon Downham, Brandon,
Suffolk IP27 OTJ
0842 814227

13

14 Glossary and abbreviations

14.1 *Abbreviations used in the text*

14.1.1 **Species**

	Common name	Latin name
CP	Corsican pine	*Pinus nigra* var. *maritima*
DF	Douglas fir	*Pseudotsuga menziesii*
EL	European larch	*Larix decidua*
GF	Grand fir	*Abies grandis*
JL	Japanese larch	*Larix kaempferi*
LC	Lawson cypress	*Chamaecyparis lawsoniana*
LP	Lodgepole pine	*Pinus contorta*
NF	Noble fir	*Abies procera*
NS	Norway spruce	*Picea abies*
OMS	Serbian spruce	*Picea omorika*
PDP	Ponderosa pine	*Pinus ponderosa*
RAP	Monterey pine	*Pinus radiata*
RC	Western red cedar	*Thuja plicata*
SP	Scots pine	*Pinus sylvestris*
SS	Sitka spruce	*Picea sitchensis*
WH	Western hemlock	*Tsuga heterophylla*

14.1.2 **Other abbreviations and symbols**

a.e.	acid equivalent
a.i.	active ingredient
cm	centimetre(s)
FC	Forestry Commission
FSC	Forestry Safety Council
g	gramme(s)
ha	hectare
HV	high volume
kg	kilogramme(s)
kPa	kilopascals
l	litre(s)
LV	low volume
m	metre(s)
ml	millilitre(s)

14

mm	millimetre(s)
mph	miles per hour
MV	medium volume
NRS	Northern Research Station (FC Research Division)
PSD	Pesticides Safety Division (MAFF, Harpenden)
p.s.i.	pounds per square inch
rpm	revolutions per minute
s or sec	second(s)
s.c.	suspension concentrate
ULV	ultra low volume
VLV	very low volume
VMD	volume median diameter
w.p.	wettable powder
w/v	weight per volume (weight active ingredient/unit volume product)
w/w	weight per weight (weight active ingredient/unit weight product)

14.2 Glossary of general and technical terms

Acid equivalent (a.e.). The amount of active ingredient expressed in terms of parent acid.

Active ingredient (a.i.). That part of a pesticide formulation from which the phytotoxicity (weedkilling effect) is obtained.

Additive. A herbicidally inactive material which is added to a herbicide formulation to improve its performance in any way.

Adjuvant. A herbicidally inactive material which is added to a herbicide formulation to enhance the phytotoxicity (killing effect) of the formulation.

Agitation. Continual mixing of a liquid preparation of a pesticide (usually at the stage of final dilution) by shaking or stirring.

Application method (or pattern). The arrangement of areas which receive an application of herbicide and the relationship of this pattern to the crop trees (when present). There is a close association between such a pattern and the application equipment used: the term 'application method' should strictly speaking refer to the applicator and its use as well as the application pattern produced. Sub-divisions are:

14

Band application. Herbicide applied to a strip of ground or vegetation, normally centred on a row of crop trees.

Directed application. The herbicide spray is directed to hit target weeds and to avoid the crop trees.

Guarded application. A spray where the crop trees are physically protected from direct contact with the herbicide by a guard or guards, usually attached to the applicator.

Incremental drift application. A form of application where the herbicide is sprayed as droplets small enough to be wind-assisted to their target and applied in successive overlapping bands so that a relatively even coverage of the whole area is achieved.

Spot application. Herbicide applied as individual spots to bare ground or vegetation, normally immediately around the crop tree.

Stem or cut stump treatments. Herbicide applied to individual stems or cut stumps wherever they may occur on a weeding site (not necessarily over the whole site).

Overall application. Herbicide is applied over the whole weeding site.

Applicator. A piece of equipment designed to distribute herbicide on to ground or vegetation.

Approved product. A pesticide which has been approved for use on grounds of safety and efficacy under The Control of Pesticides Regulations 1986.

Band application. See Application method.

Calibration. The process of calculation, measurement and adjustment (of parameters such as nozzle type, operating pressure, walking speed) by which the correct application rate is achieved.

Carrier. A liquid or solid material within which a pesticide is dispersed (e.g. solution or suspension) to facilitate application.

Chlorosis. Loss of green colour in plant foliage.

Coarse grasses. An imprecise term used to describe grasses of a generally tall, bulky, rank, stiff and often tussocky nature which are also relatively more resistant to grass herbicides. By contrast the so-called soft grasses are usually more susceptible to grass herbicides.

14

Contact herbicide. One that kills or injures plant tissue close to the point of contact or entry into the plant (contrast with translocated herbicide).

Controlled droplet application (CDA). Systems where droplets are generated by separating from points on a rapidly rotating disc or cage. Droplets so generated are more uniform in size than those generated through a hydraulic spray nozzle.

Diluent. The liquid added to a herbicide concentrate to increase its volume to an extent suitable for the applicator to be used.

Direct applicator. A piece of equipment which transfers liquid herbicide to a weed by direct contact with no intervening passage of droplets in air.

Directed application. See Application method.

Dormant period. The period of the year when the aerial part of a plant is not in active growth.

Emulsifiable concentrate. See Formulation.

Emulsion. A mixture in which fine globules of one liquid are dispersed in another, e.g. oil in water.

Esters and salts. Different groups of compounds derived from an organic acid. Esters are normally oil-soluble while salts are more usually water-soluble.

Flushing. The commencement of growth of a plant above ground, characterised by sap flow and swelling and bursting of buds. Flushing follows the end of dormancy and marks the beginning of the growing season.

Formulation.
   a. The process of preparing a pesticide in a form suitable for practical use either neat or after dilution.
   b. The material resulting for the above process.
      Types of formulation are:
   Emulsifiable concentrate. A concentrated solution of a herbicide and an emulsifier in an organic solvent which will form an emulsion on mixing with water.
   Granules. A free flowing dry preparation of herbicide (in a solid carrier in the form of particles within a given diameter range) which is ready for use.
   Liquid. A concentrated solution of herbicide which mixes readily with water.
   Suspension concentrate. A stable suspension of a solid herbicide in a fluid, intended for dilution before use.

ULV formulation. Herbicide in a special blend of oils
  intended for application through a rotary atomiser
  without dilution.
Wettable powder. Herbicide in a powder so formulated
  that it will form a suspension when mixed with
  water.
Herbi. Trade name of a controlled droplet applicator.
Herbicide. A chemical which can kill or damage plants.
Hormonal action. The mode of certain herbicides (e.g. 2,4-D)
  which achieve their effect by interfering with the growth
  regulatory mechanisms of the weed plant. Bending, curl-
  ing and deformation (epinasty) of shoots and leaves is a
  common symptom of such effects.
Incremental drift. See Application method.
Liquid formulation. See Formulation.
Low-volatile ester. An ester of an organic acid (e.g. 2,4-D)
  which has a sufficiently long chain of carbon atoms in
  the molecule to reduce the amount that evaporates during and after spraying to an insignificant level.
Low volume. See Volume rate.
Medium volume. See Volume rate.
Nozzle type. Nozzles for liquid herbicide applicators are
  described by the spray patterns produced:
    Fan. Spray droplets are emitted in a fan shape. Over-
      lapping of the tapered edges will produce an even
      distribution.
    Even fan. Spray droplets are emitted evenly over the
      width of the fan.
    Hollow cone. Spray droplets are emitted in the shape
      of a hollow cone which produces an annulus on the
      ground.
    Solid cone. Spray droplets are distributed over the
      whole area of the base of the cone.
    Anvil flood jet. Spray droplets are emitted in a wide
      fan by the stream of herbicide striking against a
      dispersing surface (the anvil) to produce a wide
      band pattern on the ground.
    Solid stream. Herbicide is emitted as a continuous jet
      (rather like a hosepipe) and not broken down into
      separate droplets.
    Variable. A nozzle in which the distribution of spray
      can be adjusted from a narrow jet to a wide cone
      pattern.

14

Pepperpot. A simple handheld container with holes drilled in the lid for distribution of granular herbicides.

Persistence. The length of time a herbicide remains active in the soil.

Pesticide. A generic term covering herbicides, fungicides, insecticides, and defined legally in the Food and Environment Protection Act 1985.

Photosynthetic process. The series of chemical reactions in the plant leaf by which sugar is made from carbon dioxide, water and sunlight. Some herbicides achieve their effect by interfering with one or more of these reactions.

Poisons rules. Regulations governing the labelling, storage and sale of materials listed as poisons under the Poisons Act 1972. See Section 2.9.

Post-planting. After the crop has been planted.

Pre-planting. Before the crop is planted.

Product. A formulation of a herbicide of fixed (but usually confidential) composition and of known strength (% content of the active ingredient or acid equivalent) which is commercially marketed under a particular brand name. It is the individual product which is given approval under The Control of Pesticides Regulations 1986.

Product rate. The amount (weight or volume) of active ingredient or product applied per unit area, per plant, per incision, etc. Because of the range of possible meanings, ambiguity should be avoided by quoting the appropriate units, e.g. litres of product per treated hectare.

Residual herbicide. One which remains active in the soil for a period after it has been applied, and will affect weeds growing into treated soil.

Resistant. Unaffected or undamaged by exposure to a herbicide applied at a stated rate. Usually used to describe the reaction of weed species to a herbicide.

Restocking area. An area where one forest crop has been clear felled and is being replaced by another.

Rotary atomiser. A herbicide applicator in which the herbicide liquid is broken into droplets of a more or less uniform diameter by being thrown from the edge of a spinning disc.

Selective herbicide. One which, if used appropriately, will kill or damage some plant species while leaving others unaffected.

14

Senescence. The annual aging process by which each autumn the leaves or above ground parts of plants wither and die back.

Sensitive. Easily damaged by herbicide.

Setting bud. The formation of buds in readiness for winter dormancy. Usually follows closely after the season's rapid growth and the development of a waxy cuticle.

Soft (or fine) grasses. See Coarse grasses.

Soil-acting herbicide. One which is active through the soil, usually entering plants through the roots.

Spot application. See Application method.

Stem treatment. See Application method.

Surfactant (or surface acting agent). A substance which is added to a spray solution to reduce the surface tension of the liquid and increase the emulsifying, spreading and wetting properties, so enhancing the tenacity of the herbicide on the treated plant.

Susceptible. Readily controlled by a herbicide applied at a stated rate.

Suspension. Particles dispersed through (but not dissolved in) a liquid.

Suspension concentrate. See Formulation.

Swath(e). A strip of ground or vegetation of a given width which receives herbicide from a single pass of an applicator.

Tolerant. Unaffected or undamaged by exposure to a herbicide. Usually used to describe the reaction of crop trees to a selective herbicide.

Total herbicide. A herbicide used in such a way as to kill all vegetation.

Toxicity. The capacity of a material to produce any noxious effect—reversible or irreversible—on the subject referred to.

Translocated herbicide. One which is moved within the plant and can affect parts of the plant remote from the point of application.

Treated area. The area of ground or plantation that is actually covered with herbicide (usually expressed as treated hectares).

Ultra low volume. See Volume rate.

Very low volume. See Volume rate.

14

Volume median diameter. The diameter in a droplet spectrum at which half the volume of the spray is contained in smaller and half in larger droplets.

Volume rate. The amount of spray solution (diluent plus herbicide) applied per unit area. Volume rates are frequently described as high, medium, low, very low or ultra low volume but several conventions exist as to the range of rates to which each term refers. The definitions used in this publication are to be found in Section 11.2 but where precision is important it is advisable to state the exact rate.

Weed spectrum. The range of undesirable species which are killed or adequately controlled by a herbicide.

Wettable powder. See Formulation.

Wetter or wetting agent. A surfactant (q.v.).

Weight/volume (w/v). A means of expressing the amount of active ingredient in a commercial formulation by relating this amount, by weight, to the volume of the formulation. In expressing the ratio as a percentage, the assumption is made that 1.0 litre of every formulation weighs 1.0 kg (e.g. 20% w/v=0.2 kg in every 1.0 litre of formulation).

Weight/weight (w/w). A means of expressing the amount of active ingredient in a commercial formulation by relating this amount, weight by weight, to the weight of the formulation (e.g. 20% w/w=0.2 kg in every 1.0 kg of formulation).

14

15 Index to weeds and chemicals

15

15

15
Printed in the United Kingdom for Her Majesty's Stationery Office
Dd291258 5/89 C60 G443 10170